PRAISE FOR

THE EDGE OF PHYSICS

"Whether he's in an abandoned iron mine in Minnesota's Mesabi Range or the frigid Siberian expanse of Lake Baikal, [Ananthaswamy] finds intrepid physicists and explains to us why these weird places are the only locations on the planet where these experiments could be done . . . Through it all he displays a writer's touch for the fascinating detail." —*Washington Post*

"[Ananthaswamy] smoothly weaves together the stories of people who help push science forward, from principal investigators to research institute gardeners, with exquisitely clear explanations of the questions they hope to solve—and why some research can be done only at the edge of the world." —*Science News*

"The ultimate physics-adventure travelogue . . . As an adventure story and a fly-on-the-wall account of remote places that most of us will never visit, *The Edge of Physics* is brilliant." —*Physics World*

"An accomplished and timely overview of modern cosmology and particle astrophysics. Ananthaswamy's characterizations of the many physicists he meets are on the mark . . . [He] conveys that cutting-edge science is a human endeavor." —*Nature*

"The author mucks in with scientists performing the world's most extreme experiments, creating a travelogue that celebrates the blood, sweat, and tears that drive our understanding of the universe." —*Guardian* (UK)

"Anil Ananthaswamy takes us on a thrilling ride around the globe and around the cosmos, to reveal the real work that goes into understanding our universe."

—Sean Carroll, author of *From Eternity to Here*

"An excellent book. The author has a great knack for making difficult subjects comprehensible. I thoroughly enjoyed it."

—Sir Patrick Moore, former president of the British Astronomical Association and presenter of the BBC's *The Sky at Night*

"A grand tour of experiments that have been built to unravel the secrets of the universe." —*Nature Physics*

"Ananthaswamy presents a compelling human interest story with some mind-blowing science attached . . . He lets everyone involved in the greatest human enterprise—discovering why we exist—express themselves. Ananthaswamy gives the reader reasons to care about scientists and their endeavors."

—The Plainspoken Scientist, blog of the American Geophysical Union

"[*The Edge of Physics* is] a smooth, calm surface: one that reflects, with wit, thoughtfulness, and humanity, a hidden world."

—*Telegraph* (UK)

"Ananthaswamy's juxtaposition of extreme travel and extreme science offers a genuinely novel route into the story of modern cosmology . . . A well-written and enormously accessible account of what it takes to push past the edge of human knowledge."

—Thomas Levenson, author of *Newton and the Counterfeiter* and *Einstein in Berlin*

THE EDGE OF PHYSICS

Anil Ananthaswamy

THE EDGE OF
PHYSICS

*A Journey to Earth's Extremes
to Unlock the Secrets of the Universe*

MARINER BOOKS | HOUGHTON MIFFLIN HARCOURT
BOSTON • NEW YORK

First Mariner Books edition 2011

Copyright © 2010 by Anil Ananthaswamy

ALL RIGHTS RESERVED

For information about permission to reproduce selections from
this book, write to Permissions, Houghton Mifflin Harcourt Publishing
Company, 215 Park Avenue South, New York, New York 10003.

www.hmhbooks.com

Library of Congress Cataloging-in-Publication Data
Ananthaswamy, Anil.
  The edge of physics : a journey to Earth's extremes to unlock the
secrets of the universe / Anil Ananthaswamy.
     p. cm.
  Includes bibliographical references and index.
  ISBN 978-0-618-88468-1
  ISBN 978-0-547-39452-7 (pbk.)
  1. Physics—Popular works. I. Title.
  QC24.5.A53 2010   530—dc22

Book design by Melissa Lotfy

Printed in the United States of America

DOC 10 9 8 7 6 5 4 3 2 1

Portions of the prologue and chapters 6, 8, 9, and 10 appeared in *New Scientist*
magazine and on NewScientist.com in slightly different form.

To my parents, Shantha and Narayana Iyer Ananthaswamy

CONTENTS

LIST OF ILLUSTRATIONS

All photographs are by the author except as noted.

AUTHOR'S NOTE

This book is more narrative than pedagogical, so concepts in physics are introduced and explained when needed. To aid readers, and prevent a lot of flipping back and forth, I've included two appendices summarizing the standard model of particle physics and the standard model of cosmology (from the big bang to the universe as it is today).

Readers will also notice that I mix up my units for physical quantities, such as length, distance, weight, and volume; for instance, sometimes the height of a mountain is given in meters and sometimes in feet. This is done purely for readability and effect. "A mile-high mountain" just sounds better than "a 1.6-kilometer-high peak"; similarly, a 1,000-foot-high balloon is more dramatic than one 305 meters high. Sometimes the metric system wins: The 27-kilometer-long tunnel housing the Large Hadron Collider near Geneva, Switzerland, would be jarring at 16.777 miles long. The use of different units also reflects the diversity of the places, people, and experiments discussed in this book.

But, after all, who knows, and who can say
whence it all came, and how creation happened?
The gods themselves are later than creation,
so who knows truly whence it has arisen?

Whence all creation had its origin,
he, whether he fashioned it or whether he did not,
he, who surveys it all from highest heaven,
he knows—or maybe even he does not know.

— FROM THE RIGVEDA 10.129, CIRCA 1500 B.C.

PROLOGUE

IT WAS THE DAY after Christmas in 2004, a bright winter's day in Berkeley, California. I was outside a café at the corner of Shattuck and Cedar, waiting for Saul Perlmutter, an astrophysicist at the University of California. The campus is nestled at the base of wooded hills that rise steeply from the city's edge. About 1,000 feet up in the hills is the Lawrence Berkeley National Laboratory (LBNL). In the 1990s, the UC campus and LBNL housed several members of two teams of astronomers that simultaneously but independently discovered something that caused ripples of astonishment, even alarm. Our universe, it seems, is being blown apart.

Perlmutter was the leader of one of those teams. His enthusiastic, wide-eyed gaze, enhanced by enormous glasses, along with a forehead made larger by a receding hairline, reminded me of Woody Allen. But what he had found was no laughing matter. In fact, Perlmutter admitted that their discovery had thrown cosmology into crisis. The studies of distant supernovae by the two teams had shown that the expansion of the universe, first observed by Edwin Hubble in 1929, was accelerating—not, as many had predicted, slowing down. It was as if some mysterious energy were creating a repulsive force to counter gravity. Unsure as to its exact nature, cosmologists call it dark energy. More important, it seems to constitute nearly three-quarters of the total matter and energy in the universe.

Dark energy is the latest and most daunting puzzle to confront cosmologists, adding to another mystery that has haunted them for decades: dark matter. Nearly 90 percent of the mass of galaxies seems to be made of matter that is unknown and unseen. We know it must be there, for without its gravitational pull the galaxies would have disintegrated. Perlmutter pointed out that cosmologists in particular, and physicists in general, are now faced with the stark reality that roughly 96 percent of the universe cannot be explained with the theories at hand. All our efforts to understand the material world have illuminated only a tiny fraction of the cosmos.

And there are other mysteries. What is the origin of mass? What happened to the antimatter that should have been produced along with matter after the big bang? After almost a century of spectacular success at explaining our world using the twin pillars of modern physics—quantum mechanics and Einstein's general theory of relativity—physicists have reached a plateau of sorts. As Perlmutter put it, he and others are now looking to climb a steep stairway toward a new understanding of the universe, with only a foggy idea of what awaits them at the top.

Part of this seemingly superhuman effort will involve reconciling quantum mechanics with general relativity into a theory of quantum gravity. In situations where the two domains collide—where overwhelming gravity meets microscopic volumes, such as in black holes or in a big bang—the theories don't work well together. In fact, they fail miserably. One of the most ambitious attempts to bring them together is string theory, an edifice of incredible mathematical complexity. Its most ardent proponents hope that it will lead us not just to quantum gravity but to a theory of everything, allowing us to describe every aspect of the universe with a few elegant equations. But the discovery of dark energy and recent developments in string theory itself have conspired to confound. On yet another winter's day in the Bay Area, more than two years after meeting Perlmutter, I got a taste of just how grave things had gotten in physics.

It was a late February afternoon in 2007. A conference room on the ballroom level of the San Francisco Hilton was filled to capacity for this session at the annual meeting of the American Association for the Advancement of Science (AAAS). Three physicists were arguing about dark energy and how it relates to some of the most serious questions one can ask: Why is our universe the way it is? Is it fine-tuned for the existence of life? Dark energy, it turns out, is not merely mysterious; it seems to be at about the right value for the formation of stars and galaxies. "The great mystery is not why there is dark energy. The great mystery is why there is so little of it," Leonard Susskind, Felix Bloch Professor of Theoretical Physics at Stanford and co-inventor of string theory, told the audience at the Hilton. He continued in a poetic vein: "The fact that we are just on the knife edge of existence, [that] if dark energy were very much bigger we wouldn't be here, *that's* the mystery."

The hope until recently had been that string theory would explain this, that dark energy's value would fall out naturally as a solution to the theory's equations—as would the answers to other puzzling questions. Why does the proton weigh almost two thousand times more than the electron? Why is gravity so much weaker than the electromagnetic force? Essentially, why do the fundamental constants of nature have the values they do? The question of dark energy is emblematic of such concerns. Nothing in the laws of physics can explain why many aspects of our universe are what they are. They seem to be extraordinarily fine-tuned to produce a universe capable of supporting life—a fact that bothers physicists no end.

But string theory's hoped-for denouement is nowhere in sight. Indeed, some physicists are slowly abandoning the notion that everything about the universe can be reduced to a handful of equations. In San Francisco, Susskind rose to address this issue. His talk was titled "Why the Rats Are Fleeing the Ship." However, abandoning reductionism hasn't meant abandoning string theory. Quite the contrary. For Susskind and many others, it has meant embracing the theory in all its mathematical glory, despite its mind-boggling consequences. One of the most outlandish implications of string theory, as it stands today, is the existence of a multiverse. The idea is that our universe is just one of a possible 10^{500} universes, if not more. And in this extraordinary scenario lies an answer to the conundrum of why dark energy and other fundamental constants have the values they do. In a multiverse, all values of dark energy and fundamental constants are possible; in fact, the laws of physics can differ from universe to universe. To explain our universe, physicists don't have to resort to tweaking and fine-tuning. If a multiverse exists, then there is a finite probability, however small, that our universe randomly emerged with the properties it has. The laws governing it give rise to stars and galaxies—and, indeed, planets and intelligent life, including physicists asking the question: Why is the universe the way it is?

This is the so-called anthropic principle, which, loosely stated, says that our universe is what it is because we are here to say so, and if it were any different we wouldn't exist to inquire. The idea is viewed by many as a cop-out, for then physicists don't have to work so hard to explain all things from first principles. Another speaker, cosmologist Andrei Linde, Susskind's colleague at Stanford, recalled his efforts to talk about the anthropic principle to physicists at Fermilab, outside Chicago, nearly twenty years ago. Linde had been warned that eggs were thrown at people who talked about such things, so he began by discussing something else entirely and switched topics midway, on the assumption that the Fermilabbers wouldn't "have enough time to go to Safeway and buy eggs."

Given string theory's support for a multiverse, the anthropic principle is gaining traction. But string theory itself is so far from being experimentally verified that many physicists find it difficult, if not impossible, to take its implications seriously. The third participant that afternoon, cosmologist Lawrence Krauss, then of Case Western Reserve University, summed up the argument for the opposition. "I think you can imagine a theory where the multiverse would be science. If one had a theory, a real theory, a real theory that predicted lots of things we see about the universe, predicted lots of things we could test, but also predicted lots of things we couldn't test, then I think most of us would say we believe the things we cannot test [such as the existence of a multiverse]," he said.

Susskind was staring daggers at Krauss by then. But Susskind's somber tone at the end of the session suggested that it wasn't going to be easy to answer critics. "All I can say is that we worry about this," he said. "[String theory] is the biggest question in physics right now. Can we make observational science out of it?"

One thing all three speakers agreed on: Only experiments could break this impasse.

The greatest advances in physics have come when theory has moved in near-lockstep with experiment. Sometimes the theory has come first and sometimes it's the other way around. For instance, it was an experiment performed in 1887 by Albert Michelson and Edward Morley—showing that the speed of light is independent of the motion of the observer—that influenced Einstein's 1905 formulation of the special theory of relativity. A decade later, Einstein produced the general theory of relativity, but it was only after experiments in 1919 verified its fascinating implication—the bending of starlight by the sun's gravity—that the theory gained widespread acceptance. And throughout the early to mid-1900s, theorists and experimentalists jostled and outdid each other as they shaped quantum mechanics. An equally fruitful collaboration occurred in the 1960s and 1970s, when particle physicists theorized about the fundamental particles and forces that make up the material world and experiments confirmed their startlingly accurate predictions. But this energetic interplay is now deadlocked. The discovery of dark energy and dark matter, along with the failure, so far, of experiments to find the Higgs boson (thought to give elementary particles their mass), has allowed theorists free rein. Ideas abound, adrift in a sea of speculation. Can the next generation of experiments in cosmology and particle physics help anchor the theories to reality?

This book is my attempt at an answer. It is a quest that took me from London, where I lived and worked, to the distant reaches of the Earth, from desolate deserts to the depths of derelict mines, from mountaintops to the bottom of the world, looking for cutting-edge experiments that promise to drag physics out of its theoretical morass. Many of the experiments I visited are tackling, each in its own way, the twin mysteries of dark matter and dark energy. But I also went to see the telescopes and detectors that are searching for antimatter, the Higgs boson, and neutrinos, which are elusive subatomic particles pervading the universe. Neutrinos barely interact with matter and travel unhindered through space, carrying information about the distant reaches of the cosmos in ways that no other particle can.

All these experiments are building the steps of Perlmutter's metaphorical stairway. My journey, too, became a metaphor: for the for-
ays that scientists are making to the very limits of their understanding—to the edge of physics.

The story begins with a pilgrimage to the 100-inch telescope at
Mount Wilson in California, where Hubble discovered that our universe is expanding, thus laying the observational foundation for the
big-bang theory and modern cosmology. The 100-inch pushed the
technological boundaries of its time, but it has long been outstripped
by modern telescopes now scanning the night skies. Every evening,
they open their giant domes to peer more than halfway across the
universe, gathering light, sometimes one photon at a time. The instruments that analyze this light are equally powerful, such as the
8.6-ton spectrograph that's helping astronomers study the universe
slice by slice with incredible accuracy. In contrast are the small,
hockey-puck-size silicon and germanium detectors, so exquisitely engineered that they are handled like works of art. They wait patiently,
day after day, week after week, for the merest hint of dark matter.

These experiments are dwarfed by gigantic balloons that soar
into the stratosphere bearing experiments that search for primordial
antimatter and study the cosmic microwave background (a radiation
left over from the big bang).

Experimental physics reaches its pinnacle at the Large Hadron
Collider, the world's largest particle smasher. Machines weighing
thousands of tons monitor the paths of subatomic particles with mi-
crometer precision. These particles spew forth from collisions of proton beams—each beam carrying as much energy as a 400-ton train
going 150 kilometers per hour. Superconducting magnets that are
colder than deep space strain to keep these beams confined to their
paths around a 27-kilometer-long underground tunnel. New particles that emerge from the cauldron of proton smashups may contain
anything from the Higgs boson to dark matter to the first hints of
extra dimensions.

These magnificent telescopes and detectors can work only in the
most extreme settings. Their surreal environments are the unsung

characters in this unfolding story—venues rarely appreciated and often overlooked. The cold, dry air above the Atacama Desert high in the Chilean Andes, where not a blade of grass can grow, allows starlight that has traveled for billions of years to enter a telescope without being smudged in its final approach by something as mundane as water vapor. (Space-based instruments, of course—such as the Hubble Space Telescope—don't have to contend with the atmosphere's deleterious effect on light.) The crystalline clarity of Lake Baikal in Siberia is crucial to a pioneering underwater neutrino telescope, and Russian physicists endure the piercing cold to camp on the frozen lake and work on their submerged instrument.

Descending into the Earth's crust affords similar benefits. Deep within an abandoned iron mine in Minnesota, physicists hunt for dark matter, their detectors shielded from the chaos of cosmic rays by a half mile of rock. The sweat-drenched miners who dug these mines with nothing more than drill bits and sledgehammers could hardly have imagined the role their mine now plays in deciphering the nature of our universe. Meanwhile, a vast and arid land in the interior of South Africa—a desolate expanse devoid of pollution—has been proposed as the site of the world's biggest radio telescope, its three thousand antennas capable of sweeping across vast swaths of the universe faster than any instrument ever built.

As extreme destinations go, there are few that compare with Antarctica, on average the coldest, driest, and highest continent on Earth. It's a land so frigid that a sharp intake of breath can sear one's lungs. Moist exhaled air freezes in an instant, and mortal danger, in the form of snow-covered crevasses, is only a moment of distraction away. Still, cosmologists cherish the Antarctic Plateau for its thin, dry, stable, and unpolluted air, and they are building gigantic telescopes to probe the cosmic microwave background with a precision that's impossible to emulate almost anywhere else on Earth. But it's not just the air above Antarctica that attracts the scientists. They are also turning the kilometers-thick ice at the South Pole into a neutrino detector. Nowhere else does there exist a block of material so massive, clear, and solid that it can be used to study the slipperiest particle in the universe. A frozen wasteland could lead us toward the correct theory of quantum gravity.

This book is a paean to the remote regions that are the soul of today's experimental cosmology. They astonish with their eloquence, whether it's the Milky Way strewn across a dark Chilean sky or the ethereal Hanle Valley ensconced in a secluded corner of the Tibetan Plateau, shielded from the world by the 8,000-meter peaks of the Greater Himalayas. Despite their differences, these places share a profound minimalism: There is nothing extraneous, none of the noise or distractions of modern society. A glaciologist I met in Antarctica spoke of the "absolute stillness" he felt on that continent, faced with only the elements, which were too extreme to ignore. Cosmology needs these places if it is to solve the pressing questions of our existence.

1 · MONKS
AND ASTRONOMERS

ON DECEMBER 31, 1930, the passenger ship SS *Belgenland* steamed into San Diego, California. One of the passengers, Albert Einstein, had come all the way from Europe. He and his wife Elsa would probably have preferred to slip into California unnoticed, but the San Diegans had other ideas. A U.S. Navy band played Christmas carols to welcome them. Five hundred schoolgirls sang songs. Reporters and photographers jostled one another to climb aboard; some even fell off the ladder. One of them asked Einstein, who was on his way to take up a winter fellowship at the California Institute of Technology, in Pasadena, whether he would look through the world's biggest telescope, atop Mount Wilson. "He is not an experimental observer," Elsa interjected. "He is a theorist."

But even the world's greatest theorist was aware of a perplexing portrait of the universe that was emerging thanks to the telescope on Mount Wilson. This new view of the universe was at odds with his own ideas. It's a testament to Einstein's scientific integrity that it took him just a few months in Pasadena, and a visit to the observatory, to accept the radical worldview wrought by the new telescope. Only the Copernican revolution, in which the sun replaced Earth as the center of the solar system, rivals its significance. And maybe not even: What happened on Mount Wilson was extragalactic in scale, with arguably greater repercussions.

Einstein might never have come to Mount Wilson if not for a man who went looking for timber in the mountains near Pasadena to make wine casks. Pasadena is at the base of the San Gabriel Mountains, not far from the Mission San Gabriel Arcángel, which in 1771 became one of the first Spanish missions to be established in California. To gather the wood to build the mission, loggers braved the nearly impenetrable canyons of the mountain range, hacked trails, and harvested timber—mainly pine and cedar. The quality of the wood so impressed Benjamin Wilson—who was elected mayor of the village of Los Angeles in 1851—that he wanted it for making wine casks for his vineyard. His workers widened an old Native American trail that started at the base of the San Gabriels and went up a cleft called Little Santa Anita Canyon. By 1864, they had established a trail all the way to a peak that overlooked the San Gabriel Valley. This peak was named after Wilson. Unfortunately, the wood proved a poor substitute for the sturdier oak traditionally used to make casks. Wilson never harvested the timber, and the trail fell into disuse. But in reaching the summit, he had opened up the mile-high mountain to an enterprise he could never have imagined.

The conversion of Mount Wilson from a potential source of lumber to a pioneering astronomical observatory began in the winter of 1903–04, when George Ellery Hale arrived in Pasadena, then a sleepy village replete with orange groves and vineyards. As director of the Yerkes Observatory on the shores of Lake Geneva, Wisconsin, Hale had been scouting for new places to build his next big telescope. The Yerkes Observatory, despite being home to a pathbreaking 40-inch telescope, was no place for serious astronomy, with its blustery winds, severe winters, and cloudy skies. Southern California's benign weather and clear skies beckoned Hale. That winter he clambered up to the summit of Mount Wilson, where he found a derelict building made of cedar logs. At night, as he lay down on a cot and gazed through the broken roof at the sky, he knew he had found the location for a future observatory.

The log building he had stumbled upon was an abandoned mountain resort called the Casino. It suited Hale's gambling instincts just fine. The Casino became the temporary base for Hale and his colleagues as they set about building the observatory. Hale cajoled his benefactors for funds, including the industrialist and philanthropist Andrew Carnegie. The summit of Mount Wilson became a place of frenetic activity. At the time, the main route to the top was a 4-foot-wide unpaved road that went up the mountain's steep southern flank, an improvement over the trail that had been hacked out by Mayor Wilson's men. Pack animals—burros and mules—were the only means of transport. This did not deter Hale. Every piece of equipment was slowly and painstakingly carried to the summit. Over time, two solar telescopes for daytime observations were built, followed by the huge nighttime telescopes—one with a 60-inch mirror and another with a 100-inch mirror. Hale risked more than his reputation, sometimes funding the enterprise with personal wealth (thanks to his industrialist father). He also took a chance by commissioning the 100-inch well before the 60-inch had been completed and its working verified. His gambles, which were in keeping with his belief that "he who would launch great ships must live in deep waters," would more than pay off.

The Casino was soon replaced by a permanent building called the Monastery, a nod to one of Hale's interests. He had been inspired by stories of monks who sought solitude in monasteries perched high among the craggy peaks of the Levant in the eastern Mediterranean. Hale saw the Monastery as a place of repose for astronomers, a place where they could study the universe undisturbed. To this end, he barred women astronomers from the observatory, afraid they would be a distraction. The Monastery had rooms for the observatory staff, a common room, library, and dining area. Today, a framed copy of a dictum of Saint Benedict welcomes any pilgrim who has traveled long distances, asking only that the visitor be content with the living conditions and that he refrain from overt lavishness. The dictum lays down a few more rules of residency, including the right of the Monastery to ask those who flout them to leave. For the offending visitor who refuses to leave, Saint Benedict sanctions a kinder, gentler version of the nightclub bouncer: "If he does not go, let two stout monks, in the name of God, explain the matter to him."

The night the building was ready for use, Hale and his fellow astronomers entered with lighted candles, as if readying for a religious ritual. They lit a fire in the granite fireplace of the common room and sat down for an all-night discussion—a start to the "elite fellowship" of brilliant astronomers interested in everything from "the specifics of optics, photography, mechanical engineering" to "the profound question of where the universe came from and how it evolved."

Astronomers were at a crossroads in the early 1900s, struggling to tease apart what they were seeing in the skies. The problem was clearly laid out in a lecture in March 1917 by Heber Curtis, an astronomer at the Lick Observatory in California. Speaking in San Francisco, Curtis talked about the difficulty understanding a certain type of nebula (from the Latin for "cloud"). These faint luminous clouds were different from the diffuse nebulae seen near the center of the Milky Way, which were clearly regions where stars were forming. They were also distinct from planetary nebulae, which were much smaller and brighter regions of gas around a dying star. The third type, the subject of Curtis's address, had been named spiral nebulae, because in long-exposure photographs they displayed a definite spiral structure, sometimes with multiple whorls of staggering complexity.

Today, majestic images of spiral galaxies are very familiar, and it is hard to imagine a time when astronomers did not know what they were. The universe in the early 1900s consisted of the Milky Way; nothing existed beyond it. No wonder, then, that most astronomers thought the spiral nebulae belonged to our own galaxy. But there were some who suspected that the nebulae lay well beyond and used the term "island universes" to describe them. Also, the spectrum of light from these nebulae suggested that they were made not merely of gas but of great clusters of stars. Unfortunately, no telescope in existence could see well enough to resolve these putative stars. And even if one could, measuring the distance to them remained a distant dream.

Summing up his talk, Curtis perfectly captured the fascination that characterized the astronomy of the time: "It is certainly a wonderful, a brain-staggering conception . . . that our own stellar universe may be but one of hundreds of thousands of similar universes . . . Familiarity with these mighty concepts most certainly does not breed contempt, does not dull our awe at the mightiness of the universe in which we play so small a part. It is very doubtful if any of those who are seriously studying the heavens ever lose their feeling of reverence for this supremely wonderful universe and for Whoever or Whatever must be behind it all."

Even as many pondered this question, Hale set about building the telescope that would help answer it: the 100-inch. When it saw first light in 1917, it was the largest telescope in the world. It remained so until 1948, a reign of thirty-one years that is simply unimaginable for any telescope today.

In 1895, two decades before the 100-inch went into operation, the American astronomer Percival Lowell decided to install a brand-new 24-inch refracting telescope—the kind made popular by Galileo—at his Lowell Observatory, in Flagstaff, Arizona. A refracting telescope, also called a refractor, is a long tube at one end of which is an "objective" lens that focuses the starlight to a point. At the other end, another lens, called the eyepiece, magnifies the image created by the objective lens. The 24-inch at Lowell (the inches refer to the diameter of the objective lens) was going to be the world's biggest. But first, the observatory had to return an 18-inch refractor they had borrowed from John Brashear, an astronomer and master optician who lived and worked in Pittsburgh, Pennsylvania.

The borrowed telescope was sent back, though a well-intentioned effort by Lowell's colleagues to clean the lens with alcohol before returning the instrument had resulted in a "faint circular mark on the

glass." Lowell sent a check of $400 to cover the damage but Bra-
shear declined the money and reportedly tore up the check. (This
was in keeping with his reputation. To his townsfolk, Brashear was
simply "Uncle John," an endearing man whose epitaph reads: "We
have loved the stars too fondly to be fearful of the night.") Intent on
compensating Brashear for the inadvertent damage, Lowell ordered
one of the optician's state-of-the-art spectrographs, which cost many
times more than $400. This spectrograph would prove crucial to an
astounding observation.

The basic idea behind a spectrograph is startlingly simple. When
the light from a star is broken down into its constituent colors by,
say, a prism, dark lines appear in that spectrum. These "Fraunhofer
lines," named after the German physicist Joseph von Fraunhofer,
who pioneered their study, are gaps, or missing wavelengths of light,
in what is otherwise a continuous spectrum. These gaps exist because elements in the star's atmosphere, such as iron, absorb light at
characteristic wavelengths. A spectrograph of a star is like a bar code
that identifies elements in its atmosphere.

The dark lines, however, are not always at the same place on the
bar code. They may all be shifted to one end (say, toward the red wavelengths) or the other (toward the blue). This phenomenon is known as
the Doppler effect. Imagine a star moving away from us. The light emitted by the star takes progressively longer to reach us, and this has the effect of stretching the wavelength of light. As the wavelength gets longer,
it starts looking redder (the spectrum of visible light is reddish at long
wavelengths and bluish at shorter wavelengths). So a star that is moving away from us looks redder than it would if it were stationary with
respect to Earth. It displays what is called a redshift. (On the flip side,
a star moving toward us causes the light to blueshift.) A star's motion
causes its bar code, or Fraunhofer lines, to move en masse toward either
the red or the blue end of the spectrum. The exact amount that these
lines are shifted is directly related to the star's velocity—or to the velocity of a galaxy or even clusters of galaxies. A spectrograph, which creates
images of a star's or galaxy's bar code, allows astronomers to measure
such velocities.

While the Brashear spectrograph was being installed at the Lowell Observatory, a young astronomer named Vesto Melvin Slipher joined the team, and Lowell tapped him to study the spectra of spiral nebulae. By 1914, Slipher had finished observing many nebulae with the new spectrograph, and what he discovered sent shock waves through the astronomical community. Nearly all the spiral nebulae were moving away from us. One particular nebula—now called the Sombrero galaxy—was speeding away at a stupendous 1,100 kilometers per second. When Slipher announced at the August 1914 meeting of the American Astronomical Society in Evanston, Illinois, that eleven out of fifteen nebulae he observed were redshifted—that is, moving away from us—he received a standing ovation. It was years before another young astronomer, Edwin Hubble, would use the 100-inch on Mount Wilson to determine the distance to these nebulae, and decades before astronomers figured out the exact implications, but on that August day everyone knew they were hearing something astonishing.

Mount Wilson and the other peaks of the San Gabriel range dominate Pasadena on a clear day. Unfortunately, on the day I arrived they were completely obscured by storm clouds. This did not deter Don Nicholson, a genial man pushing ninety and my guide for a tour of the Mount Wilson Observatory. Don's father, Seth, had been a staff astronomer at Mount Wilson during Hubble's time, and Don, as a teenager, had frequently gone up the mountain with his father. The mountaintop was where the family had spent many of its Christmas vacations.

The first and only time that Don had used the 100-inch was in 1931, when he was just thirteen. His father (who discovered Sinope, the ninth satellite of Jupiter) had devised a program of observation to pin down that satellite's highly perturbed orbit. Don volunteered to help with the 100-inch, only to be told, "No way, it's too complicated." But he wasn't put off by the dismissal. "I had been up there a lot, I had seen people doing that kind of thing, and I felt I could do it," he told me. So he kept pestering his dad, who finally relented.

On a frigid February evening, father and son went up to the dome of the 100-inch, and the teenager took his place on the platform, close to the eyepiece of the instrument. The night assistant had pointed the telescope at Jupiter, and the photographic plate was loaded. Don started the exposure and then turned around to check on his dad. The elder Nicholson was nowhere to be seen. Don realized that he had walked into a neatly laid trap. "He knew it was going to be cold. So he led me right down the garden path," Don recalled. The teenager continued observing, exposing the plate for three hours, his back to the winter air pouring through the open dome. In those days, "you were glued to the eyepiece and you didn't have a down jacket. It was cold, I'm telling you, it was cold." As soon as Don stopped the exposure, as if on cue, "there was Dad, all warm. . . ."

This bit of parental skulduggery long forgiven, Don is now the observatory's most treasured volunteer guide. He picked me up at my hotel in Pasadena, and we drove along the Angeles Crest Highway, whose construction began in 1929, well after the 100-inch was completed. We broke through the lower bank of clouds, and Don pointed to a turnoff ahead—where the highway went over a notch in the hills—that would take us to the observatory. The sun-drenched scenery did not last long; we entered another layer of clouds, and this time there was no breaking through. We reached the observatory's campus still enveloped in mist, and Don parked his car next to the galley, a small building where astronomers eat their midnight meals. Ahead of us was a footbridge spanning a shallow ravine, and across the bridge, in the rainy haze, loomed the dome of the venerable 100-inch telescope.

We stopped halfway across the bridge. Mounted on the railing was a picture of Einstein taken from the same vantage point, when he visited the observatory in 1931. Don remembered the day clearly. "I was in school at the time. My folks figured that it was better for me to be at school than to meet Einstein, which I think was a mistake." We crossed the bridge and reached the dome. Inside, after we'd climbed a flight of stairs and gone through a few more doors, Don flicked some switches. A circle of very dim lights came on, and there it was—the telescope that had helped confirm that our universe was more than the Milky Way.

We walked over to the telescope tube and stood beneath it. Through an opening in the tube, we could see the bottom of the 100-inch mirror. The glass for the mirror had been cast by the Saint-Gobain Glassworks in France—a company set up by Louis XIV to make mirrors for the Hall of Mirrors in the Palace of Versailles. When the glass blank arrived in Pasadena on December 7, 1908, having traveled by ship from France to New Jersey to New Orleans, and then by train to Pasadena, Hale and fellow astronomer George Ritchey were utterly disappointed. The 13-inch-thick, 4.5-ton blank looked like it had three layers, with air bubbles trapped in them. Indeed, the amount of glass needed to make the mirror had been so great that even Saint-Gobain, the largest glassworks in the world at the time, did not have melting ovens and ladles large enough to pour all the glass at once. They had to do so in three stages, and air bubbles had formed between pourings. Hale and Ritchey rejected the blank, so Saint-Gobain tried to make a blank in one go—and failed. Left with no choice, the team at Mount Wilson decided to grind and polish the blank they had and coat it with silver. Despite their misgivings, it turned out to be a superb mirror.

The 100-inch telescope is a reflector, a type of telescope invented by Isaac Newton. Unlike the 24-inch refractor at Lowell and the 40-inch refractor at Yerkes, which use lenses to focus starlight, the 100-inch uses a concave primary mirror. One major advantage of reflectors is that they can be much bigger than refractors. In the latter, the objective lens focuses different wavelengths of light at slightly different points along the axis of the telescope, leading to an effect known as chromatic aberration. One way to minimize this is to make an objective lens with a very long focal length. But this means that as the lens increases in diameter, the telescope grows proportionately longer, eventually becoming so lengthy as to be impractical. Another way of dealing with the chromatic aberration is to mate a convex lens with a concave one, which effectively cancels out the effect. But these multiple curved surfaces have to be perfectly shaped, a process that gets extremely difficult with large lenses. Finally, the lenses can be supported only at their edges, so heavier lenses tend to sag under their own weight. These problems are so intractable that the 40-inch Yerkes refractor, which was completed in 1897, remains the largest of its kind to this day.

Reflectors solved all these problems. The 100-inch's concave primary mirror is fixed to the bottom of a long tube made of iron trusses. The light from a distant celestial object falls on the mirror, which then focuses it to a point high up on the tube. At the mirror's focus is a flat, secondary mirror, placed at a 45° angle to the axis of the tube, which deflects the focused beam out of the tube into an eyepiece. Mirrors, unlike lenses, do not suffer from chromatic aberration. Also, they have only one active surface, which can be ground to the required shape with a high degree of accuracy. And large mirrors can be supported from beneath, preventing them from deforming. But the 100-inch came with its own challenge: the position of the all-important eyepiece.

In refractors, the eyepiece is at the bottom of the telescope tube, allowing the astronomer to stand on a platform and look through the telescope. But the eyepiece in the 100-inch is at the top of the tube, and when the telescope is pointing straight up, it is nearly fifty feet off the floor. That's where an astronomer had to sit, perilously perched on a tiny "Newtonian" platform, eyes glued to the eyepiece.

Modern telescopes use computers for tracking objects in the sky, but astronomers had to operate the 100-inch manually. Rather than observing from the cozy confines of control rooms, they were exposed to the elements. Ensconced in their aerie, simultaneously peering through the eyepiece and operating the controls, astronomers had to ensure that the telescope was tracking accurately, that the dome's opening was always overhead, and that the platform was moving in lockstep with the telescope but not bumping into it. On really frigid nights, their eyelashes could freeze to the eyepiece.

Showing surprising agility, Don clambered up a ladder with hand railings on only one side. The ladder curves from the dome's periphery toward the ceiling. As the telescope scans the sky, it swings upward from horizon to zenith. The Newtonian platform has to follow the eyepiece, sweeping a giant arc. The ladder's open side allows astronomers to get on and off the platform no matter how high it is. Operating the 100-inch could not have been for the faint of heart. Hubble's successor at Mount Wilson, Allan Sandage, wrote: "This was the astronomical observing experience at its best — a dark, quiet dome, a silently moving monster telescope, and mastery of the dangerous Newtonian platform, all in the interest of collecting data of transcendental significance."

Very few mastered the 100-inch as completely as astronomer Milton Humason, Hubble's closest colleague. The science historian Gale Christianson pictured Humason working the telescope to gather spectra of fainter and fainter spiral nebulae: "His face grotesquely illuminated by red dark-vision lamps, the freezing astronomer prodded and coaxed the reluctant beast through moonless nights punctuated by staccato winds . . . If the machinery balked, which happened all too frequently, he held the image in place by forcing his shoulder against the great cannon, and occasionally climbed onto its iron frame, bending his body at painfully awkward angles for the sake of the embryonic plate steeping in light from nebulae time out of mind." Humason himself described this acrobatic maneuver as "stretch[ing] out into nothing." It brings to mind a sailor on the edge of a tilting sailboat, ropes in hand, leaning dangerously over the water to keep the boat on course.

When the 100-inch was used for gathering the spectra of stars, astronomers could work on bright, moonlit nights. But when they were using the telescope to take photographs of extremely faint objects, such as nebulae, they could do so only on dark, moonless nights. Sandage wrote, "The 'dark-Moon types' became galactic, extra-galactic, and cosmological astronomers, while the 'bright-Moon types' used spectroscopy to divine the inner workings of stars." The dichotomy continues to this day. In the 1920s, the dark-moon types, like Hubble and Humason, were asking one of the biggest cosmological questions that had ever been posed: How far away were the spiral nebulae?

Hubble and Humason were an odd couple. On our drive up to Mount Wilson, Don Nicholson uttered a most remarkable summation of

Hubble's personality. Hubble was "at least a snob," he said. A tall, good-looking Adonis, Hubble had many affectations: an English accent acquired from his days as a Rhodes scholar at Oxford; a desire to be addressed as "Major," the rank he rose to during World War I; a penchant for wearing jodhpurs and puttees; and a fondness for his ever-present briar pipe. The snobbishness extended to his socializing. Don noted that Hubble limited his interactions with the observatory staff to those times when he needed partners for bridge, which he played to alleviate the boredom of waiting out cloudy nights. Otherwise, he and his wife preferred the company of Hollywood celebrities.

Hubble arrived at Mount Wilson after finishing his Ph.D. at Yerkes Observatory. His dream of becoming an astronomer had almost been thwarted. A domineering father had pushed him to study law, which Hubble did, while always taking enough courses in physics to keep the fire burning for astronomy. After his father's death, Hubble chucked law and attended graduate school at Yerkes. "Astronomy is something like the ministry," he once said. "No one should go into it without a call. I got that unmistakable call, and I knew that even if I were second-rate or third-rate, it was astronomy that mattered." However, Hubble was nothing if not a first-rate astronomer. Humason, who was the night assistant at Mount Wilson when Hubble made his first observation there—with the 60-inch telescope in 1919—vividly recalled the scene many years later: "His tall vigorous figure, pipe in mouth, was clearly outlined against the sky. A brisk wind whipped his military trench coat around his body and occasionally blew sparks from his pipe into the darkness of the dome . . . The confidence and enthusiasm which he showed on that night were typical of the way he approached all his problems. He was sure of himself—of what he wanted to do, and of how to do it."

Hubble made an instant impression on Humason. Night assistants, who really run the telescopes at observatories, are notorious for being hard on astronomers, making sure that astronomers earn the respect of the staff. Hubble had passed the test. But the two couldn't have been less alike. Hubble's pretentious refinements contrasted sharply with Humason's earthiness. Humason enjoyed the company of the observatory's technicians and workers. An inveterate gambler, he played a mean hand in poker, and often disappeared from the mountain to watch horse races at the Santa Anita racetrack. But Hubble was impressed by Humason's skill at working the telescopes and set him on an unlikely path to fame. Having dropped out of school after eighth grade, Humason had become a pack-mule driver for the Mount Wilson Observatory in 1915. Soon he joined the janitorial staff on the mountain, and as a janitor he started helping out with the telescope operations, showing such adeptness that he was promoted to night assistant in 1919 and then to assistant astronomer. He would work with Hubble until 1953, and along the way he picked up an honorary doctorate from Sweden's Lund University, "making him history's only scientist ever to skip from eighth grade to a Ph.D." One of the duo's first tasks was to determine the distance to Andromeda, a prominent spiral nebula. But the method Hubble used was not his. Its discoverer was an unassuming woman who often went unmentioned and whose gender would have barred her from observing at Mount Wilson.

Henrietta Swan Leavitt worked as a human "computer" in the late 1800s. Her job was to count stars at the Harvard College Observatory, which had taken on the ambitious task of cataloging every star in the sky. The work demanded painstaking manual inspection of photographic plates to pin down the stars' position, color, and brightness. Edward Pickering, the director of the observatory, recruited men to do the job, only to be frustrated by their "lack of concentration and failure to pay attention to detail." Convinced that women could do a better job, he fired the men and hired a team of women, who were nicknamed "Pickering's Harem." Not only did Pickering get a more diligent team of workers, he paid them only about half as much as he paid the men. And he did not have to worry about the women wanting to make their own observations, for (as at Mount Wilson) they were not allowed to use the telescopes. It was as part of this team of desk-bound computers that Leavitt discovered something extraordinary.

She was studying photographic plates of the Small Magellanic Cloud (SMC), a nebula that was clearly visible from the Southern Hemisphere and had been photographed by Harvard's observatory at Arequipa, Peru. At the time, distances to these nebulae were unknown, and no one knew how to go about calculating them. Leavitt, as she pored over the plates, discovered twenty-five stars known as Cepheid variables, all in the SMC. Cepheid variables are stars whose brightness fluctuates periodically. Leavitt correctly surmised that the twenty-five Cepheid variables must all be at nearly the same distance from Earth, since they were all part of the same cloud. Using this assumption, she worked out a relationship between the brightness and the periodicity of these variable stars: The more time it took for a Cepheid variable star to wax and wane, the greater was its intrinsic brightness. Finally, here was a way to look at a star and gauge its intensity—you just had to measure its periodicity. Then, given the star's intrinsic brightness and also its *apparent* brightness from Earth, you could calculate its distance—for brightness falls as the square of the distance.

However, Leavitt's work allowed you to calculate only the relative distance from Earth of two Cepheid variable stars. Someone still had to figure out the distance to at least one Cepheid variable before you could figure out the absolute distance to, say, the Small Magellanic Cloud. That task was taken on by others: first, a Danish astronomer named Ejnar Hertzsprung, who triangulated the distance to a Cepheid variable star in the Milky Way by using the sun's motion through our galaxy, and then the American astronomer Harlow Shapley, who further refined this yardstick. Now astronomers could calculate the absolute brightness of a Cepheid variable by measuring the rate at which it blinked, and from that they could work out its distance. The heavens had indeed opened up to them.

In fact, so significant was this advance that Shapley thought that he, not Leavitt, should be recognized for it. In 1925, the Swedish mathematician Gösta Mittag-Leffler, when he heard of Leavitt's work, wrote to her, saying, "I feel seriously inclined to nominate you to the Nobel prize in physics for 1926." Unfortunately, the letter never reached Leavitt, who had died a few years earlier. Shapley, who was then the director of the Harvard observatory, received the letter instead and, in his reply to Mittag-Leffler, damned Leavitt with "faint praise," arguing that he had in fact interpreted Leavitt's observation, thereby suggesting that his was the real intellectual accomplishment.

In any event, both Leavitt and Shapley had been tinkering with Cepheid variables in and around the Milky Way. With the 100-inch telescope at his disposal, Hubble extended their reach, finding the first Cepheid ever to be discovered in the Andromeda nebula. Using Shapley's calibration, Hubble calculated that Andromeda was about a million light-years from us—far beyond our own galaxy, which was known to be only about 100,000 light-years across. This was the first real evidence that the spiral nebulae were indeed island universes, as Curtis had so elegantly posited in his 1917 talk.

Vesto Slipher, meanwhile, had accumulated masses of redshift data showing that many of these nebulae were receding from us. So Hubble and Humason decided to push the instruments to their limits, gathering redshift data for as many nebulae as they could—the fainter the better. There was no better telescope for the job than the 100-inch. Finally, in 1929, Hubble published his now-famous paper—"A Relation Between Distance and Radial Velocity Among Extra-Galactic Nebulae"—which was a mere six pages long. He had found a linear relationship between the distance to "extra-galactic nebulae" (to the end of his life, Hubble refused to call them galaxies) and the speeds at which they were moving away from us. The velocity of a galaxy was, within limits of observational error, proportional to its distance from Earth, and the constant of proportionality came to be called the Hubble constant. Basically, a galaxy twice as far away as another was moving twice as fast, and one thrice as far away was moving at three times the speed. Here, finally, was a simple way to find the distances to these clouds, most of them so faint that astronomers could barely discern them, let alone see individual stars in them. One just had to find their redshifts. The only other value needed was that of the Hubble constant, and Hubble used Cepheid variables to find the distances to the nearest nebulae to calculate the constant.

It was around then that Einstein arrived in Pasadena, and in February 1931 he was driven up to Mount Wilson with Hubble in a "sleek Pierce-Arrow touring car" that had been bought for the occasion. At the observatory, Einstein was suitably impressed by the telescopes, and, "to everyone's alarm, he insisted on climbing over the steel framework of the 100-inch while rattling off an extensive knowledge of its every appliance." Elsa, who was told that the telescope was being used to figure out the size and shape of the universe, reportedly retorted, "Well, my husband does that on the back of an old envelope." Despite Elsa's faith in his abilities, Einstein had turned out to be wrong on this count: His general theory of relativity suggested that the universe could not be static, but his belief in an unchanging universe was so strong that he had tweaked the equations to accommodate it. Faced with Hubble's discovery, Einstein conceded his error.

Meanwhile, Hubble's star rose beyond the heights of Mount Wilson. He became a true celebrity, feted by everyone. He was even a guest of honor at the 1937 Academy Awards. The glare of his compelling personality and fame has blinded the world to the contributions not only of Slipher but also of Carl Wirtz, dubbed the "European Hubble without a telescope," who had surmised from earlier redshift measurements that the universe was expanding. Meanwhile, Alexander Friedmann, a Russian mathematician and meteorologist, had solved Einstein's equations of general relativity for a universe that was uniform on large scales and had shown that the universe could not be static but rather had to be either expanding or contracting. And a Belgian priest-mathematician, Georges Lemaître, had used the Cepheid-variable-star data, along with Einstein's equations, to determine that the universe had to be expanding. Lemaître, in fact, predicted a linear relationship between the distance and velocity of galaxies two years before Hubble announced the discovery. A year after Lemaître, a Caltech mathematician, H. P. Robertson, independently arrived at the same conclusion.

Nonetheless, it was Hubble who shattered our view of a static universe. It took someone of his personality and persuasive skills to convince the broader astronomical community that the universe was expanding. Hubble himself refused to make that claim, preferring to report on what he was observing and leaving the theorizing to others. But his data strongly suggested that the universe had a beginning, contrary to the conventional wisdom of the day. If the galaxies were speeding away, then they must have been closer at one time, and if you worked backward, the linear distance-velocity relationship was such that there would have been a time when all the matter in the universe would have been at the same place. Our universe went from being merely the Milky Way, forever in its current form, to one in which our galaxy was one of many, all of which were speeding away from us in a manner that more than hinted at a primordial explosion. The idea of a big bang began to take shape.

Don Nicholson and I stepped out of the 100-inch dome at Mount Wilson. The clouds had thickened. All around us was evidence of a cold winter: snow piled up on the sides of roads, sometimes a few feet high. We walked across the footbridge to the galley to eat the sandwiches we had brought with us. The galley was also called the "midnight lunch shack" and that's what it was—a shack, situated midway between the 100-inch and the 60-inch telescope domes. During the observatory's early, heady decades, the midnight lunch depended on how the 100-inch was being used. When the astronomer was doing the relatively easy stellar spectroscopy, his assistant would walk down to the shack and prepare lunch, eat his share, and then go back to the telescope to relieve the astronomer, who would then go down and have his.

But the situation was entirely different if the telescope was being used in the delicate and dangerous Newtonian mode. The assistant could not leave the astronomer to prepare lunch. So both took an hour-long break and wandered over to the shack—as it happened, for not much succor. "As late as 1955, no food was stored in the midnight lunchroom," wrote Sandage. "Instead the food had to be

packed in a small basket and toted up the steep path from the Monastery each night. This responsibility fell to the 60-inch observer,
whose hike culminated in hanging the basket on a hook above the
floor of the 60-inch dome, safe from scavenging animals. At midnight, the basket would be taken to the lunch room. This was a lot
of buildup for very little payoff, for rarely was there much food in the
basket." The frugal observatory director had allotted "two pieces of
bread, two eggs, some butter and jam, and enough tea or coffee to
make a single cup."

We ate our relatively sumptuous sandwiches and drove down the
narrow road to the Monastery. Astronomers would have routinely
walked down the quiet fir-lined road after a night of observing. Don
drove slowly, worried about the snow on the road. We came to a stop
in front of the two-storey section of the Monastery, which is built on
a spur overlooking the valley below. On a clear day, the view is ap-
parently spectacular, with an uninterrupted line of sight to Catalina
Island, 22 miles offshore.

We had a quick look at the basic sleeping quarters—divided into
rooms for the daytimers and the nighttimers. When the observatory
first opened, it had only the solar telescope, and the astronomers
observed the sun during the day and slept at night. That changed
with the arrival of the optical telescopes and astronomers who had
to sleep during the day. The only sensible solution was to keep the
two groups separate, with unwritten rules to keep quiet while others
slept. But the rules were not often followed. "Both the night and day
observers kept separate blacklists identifying those who routinely
violated these edicts," wrote Sandage. On a lighter note, he added,
"Some nighttime observers were whistlers, especially after a good
night; the sleeping solar observers got to know us well."

From the sleeping quarters, we entered the common room, which
had a library of deliciously old leather-bound books—by Lord Ray-
leigh, Lord Kelvin, James Clerk Maxwell, Henri Poincaré, and, of
course, George Hale—their edges and corners frayed, the gold let-
tering dimmed with age. Then there was the infamous eleventh edition of the *Encyclopaedia Britannica*. Hubble was known to come to the

library and spend an hour reading the encyclopaedia, boning up on some obscure topic before sitting down for lunch. Thus fortified, he would proceed to the adjacent dining room and, befitting his status as the observer for the 100-inch telescope, would take his place at the head of the twelve-man table. The hierarchy at the dining table was rigid: The 60-inch observer sat to Hubble's right, the 150-foot solar tower observer to his left, and the others farther along in order of their presumed importance. Once so seated, Hubble would subject his poor colleagues to the Mount Wilson version of the "high table at Oxford." As the head of the table, he had the privilege of priming the conversation, and he would do so by interrogating others on the very subject he had just studied. "Then he could dominate the conversation," said Don. Sandage's more charitable interpretation of this behavior is that Hubble, "almost pathologically shy around colleagues with whom he had little other contact, had hit upon a foolproof way of controlling a potentially awkward social situation."

The Greater Los Angeles area, which Mount Wilson overlooks, is hardly the quiet, unpopulated, unpolluted region that Hale once found so appealing. However, there is a section of California's coastline that still remains pristine, providing a glimpse of what Hale might have encountered. These are the forested slopes of the rugged Santa Lucia Mountains, about halfway between Pasadena and San Francisco. Amid these slopes is a Camaldolese monastery, where solitude and silence are a way of life, much as Hale desired for his astronomers.

Driving north from Pasadena, I arrived at the monastery close to sundown. I was shown my room, a functional setup with a single bed, writing table, and small chest of drawers. The table overlooked a tidy garden, beyond which the land tumbled down to meet the Pacific. The view of the bay below was lavish in contrast to these spare and simple furnishings. For early European explorers, who had sailed into these waters and had yet to climb this soaring shoreline, the mountains must have looked as dramatic from the ocean as the ocean does from such heights. Juan Rodriguez Cabrillo, the first conquistador to sail into Lucia Bay, wrote in his diary in 1542 that the

Santa Lucias were "mountains which seem to reach the heavens, and the sea beats on them; sailing along close to the land, it appears as though they would fall on the ships." The description was not hyperbole. Of all the mountains along the coast of the contiguous United States, the Santa Lucia range rises the most steeply from the sea.

It took just a night for the hubbub of Southern California to recede. When I woke up the next morning, the only sounds were the ticking of the bedside alarm clock and the hiss of the gas heater. The rising sun had brushed the clouds on the horizon with the palest possible hues of pink and orange. A gibbous moon was still visible. Just outside my window, among green firs and pines, was a gnarly old tree stripped of bark, its stark white branches ending in a tangle of leafless twigs. There was little wind, and the blue-gray ocean looked immense and calm, merging into the indistinct, foggy horizon. A ship glinted in the fog as it caught the sunlight that was creeping up on the monastery.

The monastery bells reverberated through the still morning air as the sun finally rose over the mountaintops. The birds woke up, too; a woodpecker alighted on the wooden fence outside my window and started to sound a steady beat. A blue jay flew by. On the gnarled tree, either a bird of prey was waking up for the day or a rather large owl was getting ready to call it a night—I couldn't tell from the distance. Nature's sounds only served to enhance the quiet.

That night, I stepped out into the garden overlooking the ocean. I couldn't see it for the dark, the moon not having risen yet, but the muffled roar of waves and the cool breeze gave away its presence. The birds were quiet; the only creatures that seemed intent on communicating were frogs. There were no lights where I stood, and the stars were the sharpest I had seen over the past few days. Orion had moved from a position directly overhead earlier in the evening down to the western horizon. I could imagine Hubble or Humason sitting on the platform of the 100-inch, eyes to the eyepiece for hours, as the telescope tracked the nebula near Orion's belt. Standing in that tiny garden, I was deeply envious of astronomers, who had by design or fate adopted a profession that required them to be awake at night.

Three days passed in more or less complete silence. There was only the clicking of my computer keyboard and the occasional conversation with the monks. On the third day, nature closed in on the place. Fog and cloud rose from the ocean and swept over the mountains, engulfing us. With visibility down to less than a few hundred feet, even visual dialogues with the land were muted.

Earlier that day I had met Father Robert, who was manning the monastery bookstore. I told him I was traveling around the world to see telescopes. His eyes lit up. When he had been in college in Southern California as a liberal-arts major, he had chosen to study astronomy, thinking it would bring him a sense of awe and wonder. "And it did," he exclaimed. But his calling had been clear to him. "Already in high school I was thinking of the monastic life, because I did have an experience of a mysterious [nature]," he said. Whatever it was, it prompted him to seek out a community that supported contemplative prayer, so he came to the Camaldolese monastery in 1959 as a young man of twenty-one. He soon took his monastic vows for life. But something drew him back to study, and he left to earn a doctorate in theology at Fordham University, where he worked on reconciling the growing awareness of our immense, expanding universe with his religious faith. He found inspiration in Einstein's words: "To know that what is impenetrable to us really exists, manifesting itself as the highest wisdom and the most radiant beauty which our dull faculties can comprehend only in their most primitive forms—this knowledge, this feeling is at the center of true religiousness."

Father Robert returned to the monastery after completing his doctorate. "I'm delighted I didn't do anything else, like go off and become a professor," he told me. As he sees it, he has only one lifetime, and he wants to dedicate it to this ultimate mystery. "We think that there is a deep experience that [one] can't express—suddenly an amazement. People often have this, without realizing it, without giving it fancy names. I think the ocean calls out that astonishment, and certainly astronomy [does]."

When I mentioned that I had just visited the Mount Wilson Observatory, Father Robert smiled. He said that his last name was Hale and that both he and George Ellery Hale could be traced back to a Thomas Hale, who had left Hertfordshire, England, and landed in Newbury, Massachusetts, in 1637. "My family had a real sense of genealogy," he said. "I remember that we had this list of our greats. George Hale is one of [the greats]."

Cosmologists would agree. The observatory Hale built is called the birthplace of modern observational cosmology. Staff astronomers at Mount Wilson Observatory continued until the 1950s to make crucial contributions to our understanding of the universe. However, the light pollution from Los Angeles and surrounding areas worsened, and Hale, who had already secured $6 million in new funds to build a 200-inch telescope, began scouting for another mountaintop. He settled on the 5,600-foot-high Mount Palomar, about 100 miles southeast of Pasadena.

But one of the first telescopes to be built at Mount Palomar, at the insistence of Bulgarian-born astronomer Fritz Zwicky, was a much smaller, 18-inch telescope. It was a newly invented instrument known as a Schmidt telescope, which used the combination of a primary mirror and an aspherical lens to create images that were free of aberrations across its entire wide field of view. The 18-inch Schmidt could survey vast swaths of the sky much faster than a larger telescope. In the mid-1930s, Zwicky used it to open up a whole new mystery. He observed that the galaxies in the Coma cluster were moving too fast to make any sense. The average speed of galaxies within a cluster depends on the cluster's total mass, and the mass of the cluster, which Zwicky estimated from the total number of galaxies that could be observed, was too low. Theoretically, the mass did not have enough gravity to keep the galaxies bound to the cluster—they should have dispersed long ago. Yet there they were. Something was keeping them together. Far ahead of his time, Zwicky suggested that more than 90 percent of the mass of the Coma cluster was made of unseen matter. Nearly eighty years later, physicists are still trying to make sense of this "dark matter." On the trail of one such pioneering experiment, I soon found myself deep within an abandoned iron mine in Minnesota.

2 · THE EXPERIMENT THAT DETECTS NOTHING

THE AIR WAS a chilly 50°F. At 2,341 feet underground, the temperature held steady. Despite the discomfort, I was better off inside this defunct iron mine, for on the surface an already freezing Minnesota winter was being buffeted by a blizzard and was a good 25°F colder, not accounting for wind chill.

The day before, I had outdriven the storm, racing ahead of it from Minneapolis to a small mining town in the northeastern part of the state. The young woman at the airport's rental car counter had unintentionally tipped me off. "Mom, I'm gonna try and leave by noon, before the storm hits," she said into the phone.

It was 6:30 A.M.

"What storm?" I asked. I had just flown in from a rather sunny San Francisco.

"Haven't you heard? We're expecting one of the worst snowstorms of the season," she said.

Just my luck. I was in Minnesota for three days, and it had to coincide with a slushy mix of snow and sleet in what had otherwise been a relatively dry but cold winter. The storm was coming from the south. I quietly asked for a bigger, safer car than the one I had reserved and was soon headed north, leaving behind the Twin Cities of Minneapolis and St. Paul.

The rhythmic *thwack* of rubber tires hitting the seams on the road was soporific. I wanted to pull over and take a nap, tired as I was after the red-eye flight, but the thought of the approaching storm kept me going. After a few hours, the highway entered the Mesabi Range—low, rolling hills known for their iron ore. The needleleaf forests kept getting thicker, with the drooping, snow-laden spruce branches resembling giant milk-covered mustaches. The wind picked up, spraying snow onto the road in a fine mist. Oncoming cars had their headlights on, and it was only noon.

I passed the geological marker for the Laurentian Divide and sighed in relief. The divide separates the great watersheds of North America: To the north, water flows into the Hudson Bay and the Arctic Ocean; to the south, it drains into the Great Lakes and the St. Lawrence and Mississippi rivers. I had heard that bad weather from the south sometimes stays south of the divide, and as I drove across it I felt safe, as if I were sailing into a harbor.

I arrived at a lodge in the town of Tower, named after Charlemagne Tower, a politically savvy industrialist who founded the Minnesota Iron Company in 1882 and began mining the rich veins of hematite in nearby Soudan. By the time the Soudan Mine closed operations in 1962, it was the deepest mine in Minnesota. Today, it hosts one of cosmology's most sensitive experiments: the Cryogenic Dark Matter Search (CDMS).

The CDMS experiment has set the most stringent constraints yet on the nature of the unseen mass of the universe. Visible stars, gas, and dust constitute a mere fraction of the universe's total mass. The rest is "dark," emitting no light. Indeed, it barely interacts with normal matter. If dark matter and normal matter were interacting electromagnetically, then astronomers would have detected photons from such interactions. They don't see any. So a dark-matter particle—whatever it is—must pass right through normal matter, interacting only if it smashes directly into a nucleus. But because a nucleus is an insignificant part of an atom by volume, such collisions are highly unlikely. No wonder, then, that although our galaxy is thought to be awash in particles of dark matter, the Earth seemingly plows through a sea of this mysterious stuff without incident.

Nothing in the so-called standard model of particle physics—which describes the particles of normal matter and the fundamental forces that act on them—can explain dark matter. One way to solve the mystery would be to witness a dark-matter particle smashing into a nucleus of normal matter. The knowledge gleaned from even a single such unlikely event would reverberate through all of physics. But on Earth's surface, any interaction with a dark-matter particle would easily be swamped by collisions with other kinds of particles, due to everything from radioactivity to cosmic rays. So scientists have been driven underground, in search of an unnatural silence that will let them "hear" the *ping* of a dark-matter particle hitting one of their delicate detectors.

Across the world, teams have descended deep into mines. In 2003, the Americans established their dark-matter experiment in an underground physics laboratory that had been constructed two decades earlier in the abandoned Soudan Mine. Crucially for the physicists, the miners had left behind a precious resource: the mine shaft and a working elevator, once a lifeline for the miners and now for the laboratory. The elevator is nothing like the straight-up-and-down affair you might find in a modern mine. When the Soudan Mine opened, in 1884, it was an open-pit operation, but the company soon realized that the ore ran deep rather than horizontally, so the miners dug a shaft alongside a rich vein of ore that went down into the Earth. To minimize the length of horizontal tunnels needed to get to the iron, the shaft tracked the column of ore closely, so instead of going straight down, it descended at a 78° angle, paralleling the vein itself. The mining company built a gigantic steam-powered hoist to operate a pair of cages. This was later upgraded to an electric motor, which is still used today. The cages act as their own counterweights: If one is going down, the other is coming up. They run on rails to prevent them from swaying inside the shaft. Despite this ostensibly stable ride, my first trip down into the mine was unnerving.

Michael Dragowsky, a physicist from Case Western Reserve University, in Cleveland, Ohio, and a shift leader for CDMS, met me at the mine's headframe, the structure that supports the pulley for the hoist. We and a few others donned hardhats and entered the cage. The door slammed shut, the cage started rolling down, and soon everything went completely dark. The phrase "pitch-black" suddenly made sense. I could see nothing except the faint fluorescence of a sticker on someone's hardhat. The sound of the steel wheels rattling down the rails, confined as we were to the narrow shaft, was as thunderous as a jet engine's roar. I braced myself against the cage walls as best as I could. Everyone fell silent, partly because you had to shout to be heard but also because there was something about the darkness and the earsplitting noise that forced an internal stillness. In what seemed longer than the advertised three minutes, the cage reached the bottom level of the mine.

We stepped out into a dimly lit cavern, to be confronted by a sign reading "Level No 27. 2,341 feet below the surface." We had plunged nearly half a mile into the Earth. The cavern smelled of barren dustiness; nothing green had ever grown here and nothing living had ever crawled here, until humans bored their way down. Dragowsky remarked that the smell reminded him of underground parking garages. A small bat flitted past and flew into a drift—an old tunnel, on our left, that had been dug decades ago by the miners to reach the ore. Rail tracks disappeared into its depths.

Hours later, I would join Jim Essig, the park manager for the Soudan Underground Mine State Park (which runs tours of the mine), for a trip into the drift. We sat in an open-topped railcar that rumbled into the narrow tunnel, which was barely 10 feet wide and about as high. If I had stretched out my hand, I could have touched the wall as we sped past. All conversation stopped again, subjugated to the reverberating din of the railcar. Dim yellow incandescent bulbs hung from the ceiling in a long line that disappeared into the darkness ahead. The railcar finally reached the end of the three-quarter-mile-long tunnel. We alighted, climbed up a steep spiral ladder and entered a stope—a cavern left behind after the ore had been extracted. When the mine was operational, the ore would have been dropped into the drift below, loaded on railcars, rolled down the gentle slope to the elevator shaft, and hoisted to the surface. More than 17 million tons of ore were extracted from the Soudan Mine over its lifetime.

We were in a mammoth stope, which was prevented from caving in by columns of rock that had been left standing. For the benefit of tourists, there were mannequins dressed to look like miners. In the low lighting, they gave the chamber a creepy feel. Then Essig switched off the lights, and suddenly I couldn't even see the back of my hand when I brought it up to my face. He lit a candle, and our eyes slowly adjusted to its feeble light and the eerie darkness. The flickering flame cast long shadows on the cavern floor and walls. In the very early days of the mine, teams of three miners would have worked in candlelight, painstakingly chipping at the rock, one miner holding a drill bit to the rock while the others took turns hammering against the bit. In between the sledgehammer blows, the man holding the drill bit would rotate it a quarter turn, and so it went, bit by bit, eating into the steel-gray hematite. Even at the height of early mining technology, the men would have used only hand-held drills powered by compressed air. And yet some of the stopes had ceilings more than a hundred feet high—that's how much ore had been extracted in nearly a century of mining.

We took the railcar back to the shaft. Before heading to the CDMS laboratory, I sat down with Essig inside what used to be the miners' lunchroom—a low-ceilinged cavern about 10 feet across and 20 feet deep. A long wooden table and two benches flanking it were the only pieces of furniture. Jackets, gloves, and tools lay scattered about, as if the miners had just left work. Not everything here was original; some of this flotsam had been set up for tourists. However, lunchrooms on other levels remain untouched, and mostly inaccessible now, littered with fragments of girlie magazines, tins of chewing tobacco, and even parts of a Christmas tree—the artifacts of another era.

Essig escorted me to the laboratory, on the other side of the drift. We walked through a set of massive green doors into a dramatic, brightly lit cavern, 40 feet high, 50 feet wide, and a staggering 270 feet long. After the claustrophobic confines of the cage and the dingy drift, the laboratory cavern—excavated using mechanized equipment—is a modern marvel. A huge mural, about 60 feet long and 25 feet high, celebrating the history of particle physics, dominates one of the shotcrete-lined walls. In creating it, the artist Joseph Giannetti used 50 gallons of paint, several assistants, and a window-washing platform to help him navigate the massive wall. It's a striking visual: a dominant mandala in the center flames outward in shades of yellow, red, and brown, while the background stills the senses with its nebula-inspired colors of blue and purple. Hidden in the mural is the word "change" in twenty different languages. "When I started to create this image," Giannetti told an interviewer, "I was feeling something about energy, something about universal language, something that was abstract and yet so clear . . . Accepting that all things are in motion—you, me, the world, the universe—from the subatomic level to the universal level, there is then only one constant: change."

It is a fitting tribute to the morphing of the mine itself, from America's premier source of iron ore to one of its finest underground laboratories. As the lab was being set up (and before the mural was painted), Essig brought Louis Cvetan, an aging miner who had once worked the mine, to see its transformation. The normally loquacious Cvetan quietly walked up and down the length of the cavern. "When he came back, there was a dang tear in his eye," said Essig. "He was an old hardcore miner all the way, so I said, 'Louis, what's the matter? You feel OK?' And he said, 'You know, if it wasn't for what us guys did, this wouldn't be here.'"

The hard work of the miners has indeed made the extraordinary laboratory possible. The CDMS researchers, however, might never have arrived if it hadn't been for a young astronomer named Vera Rubin. In the mid-1960s—not long after the mine had shut down for good—Rubin began poring over some puzzling spectrographs of the Andromeda galaxy. Something about the universe's mass was not quite adding up.

Denying an astronomer a view, any view, should be deemed a sacrilege. Vera Rubin was understandably miffed at the one from her office window. The campus of the Carnegie Institution of Washington's Department of Terrestrial Magnetism in D.C. was ablaze with the reds and golds of thick autumn foliage—even the gray October sky and persistent drizzle could not dampen the effect. But you could barely see the tops of those trees from Rubin's office, with its absurdly high and deep windowsill. The relatively short-statured Rubin was upset when she first saw the sill, and she threatened to call the contractor and have him build her a balcony, so that she could enjoy the view. The threat fell on deaf ears. A frustrated Rubin created a surrogate view by putting up posters of flowers on the wall below the sill.

That was years ago, but "I'm still angry," she said. It was hard to imagine this silver-haired grandmother, who was about to enter her eighties, getting angry at anything. Sitting in her office, an entire wall of which was lined with observing logs from five decades of astronomy, she spoke gently, her voice quivering occasionally. Her career as an astronomer had been nothing short of pathbreaking, starting with her 1951 master's thesis at Cornell University, where she worked on the then controversial topic of residual motions of galaxies. Did galaxies have any motions apart from those caused by the expansion of the universe? Her answer was yes. The work attracted a fair amount of publicity, much of it negative. But George Gamow heard about it. "He started calling me and asking me questions about astronomy, and they were all brilliant questions," Rubin recalled.

Getting noticed by Gamow was heady for the young Rubin, for Gamow was no ordinary physicist. In 1934, he had arrived at George Washington University, in D.C., as a "6-foot-3-inch, 30-year-old, flaxen-haired, Ukrainian émigré scientist [with] startlingly blue eyes [that] twinkled myopically behind lenses that resembled the bottoms of cider bottles." Gamow had defected to the United States from the Soviet Union with his wife, on the pretext of attending an international conference. This after the duo had failed an earlier escape attempt by paddling across the Black Sea in a kayak. "The first day was a complete success," Gamow wrote later about this episode. He recalled marveling at the sight of a porpoise silhouetted against the setting sun. But the next day Gamow and his wife were beaten back by the wind, and instead of landing in Turkey, they found themselves back in the Soviet Union.

By the time Gamow eventually defected to the United States, he was already famous in Europe. He soon gained further fame in America. Most important, he came to be firmly associated with the big-bang model of the universe. Rubin was understandably excited about the opportunity to work with him; she moved to D.C., studied for her doctorate at Georgetown University, and began meeting with Gamow at the Department of Terrestrial Magnetism.

Rubin fell in love with the DTM campus, and one day, now a full-fledged Ph.D., she walked in and asked for a job. Hiring a woman astronomer would be a first for the department, which had been founded in 1904 to map the Earth's geomagnetic field. Its observers trekked through inhospitable regions and circumnavigated the globe on the *Carnegie*, a ship made mostly of nonmagnetic materials. Naval tradition had kept the DTM a male bastion. In 1965, when Rubin arrived, there were no women on the faculty, and there was a time when even the secretaries were men. But the department hired her as its first female scientist.

Soon after arriving, Rubin nailed the idea that galaxies have much more mass than can be accounted for by luminous matter alone. She was not the first to grapple with the problem of missing mass. By the mid-1930s, Fritz Zwicky had already observed that galaxies in the Coma cluster were moving at speeds too high to be explained by the cluster's observed mass. He reasoned that there had to be more mass in this and other clusters than met the eye, providing the gravity needed to hold the clusters together. But astronomers had never quite taken Zwicky's assertion about unseen mass seriously.

Early astronomers can be forgiven for ignoring evidence of dark matter from studies of galactic clusters, given how difficult it was to observe them with the telescopes of the day. However, later studies of individual galaxies showed that even the best of scientists can be blinkered. By the 1960s, astronomers were measuring the velocities of stars in galaxies by studying their redshifts. But they would study only the inner precincts of a galaxy and then draw rotation curves—graphs that map the velocities of stars as a function of their distance from the galactic center. "What they would do is draw a Keplerian fall-off," said Rubin, referring to the application of Kepler's laws of planetary motion to galaxies. These laws say that planets closer to the sun travel faster than those farther away; when applied to galaxies, they require the speed of stars to drop with increasing distance from the galaxy's center—that is, if stars, gas, and dust are the only components of a galaxy. Astronomers simply assumed that, in accordance with Kepler's laws, the velocities of stars would fall off as they moved away from the center, and they merrily drew rotation curves to reflect that assumption.

This was the state of knowledge in 1965, when Rubin teamed up with Kent Ford, a young astronomer with prodigious practical skills. Ford had pioneered the image tube spectrograph, a multistep device in which each stage took photons from the previous stage and amplified the light into ever more photons. The instrument captured light from parts of galaxies so faint that astronomers just looking through the eyepiece could barely see anything there. Rubin and Ford made an unusual but effective team. "Kent was a magnificent instrument builder, and he knew a lot of astronomy, but once the observations were done he had no interest in the analysis," Rubin remembers. She would do the analysis and write up the paper for publication (including his name on it, "of course"), and the paper would sit on Ford's desk for weeks, sometimes months. When pressed to read it, Ford would tell her he didn't have to and that she could remove his name and publish it. She would eventually publish it and keep his name. "We made a wonderful team," she said, "because the instrumentation was really lovely for that time. We were observing things that you couldn't possibly see in the telescope."

Rubin and Ford started off by observing the extremely faint regions of the Andromeda galaxy, a spiral giant 2.5 million light-years away which dominates our galactic neighborhood. Each night, the duo collected spectra of ionized hydrogen in different regions of the galaxy, far from its bright center. The spectra would show how fast the hydrogen was moving around the galactic core. As Rubin saw the spectra emerge on the photographic plates, she was struck by how similar they all looked. The galaxy was rotating uniformly, no matter how far away from the center they got. Unlike her predecessors, Rubin accepted the data, refusing to draw a convenient Keplerian curve. "I believed very early on that this was real, and that this was telling us a lot about the universe," she said. "But I didn't know what it was."

This was 1970. No one quite knew the significance of Rubin's findings, and some astronomers remained reluctant to accept them, even though Rubin buttressed the Andromeda study with similar observations of other galaxies. Her own rotation curves were suggesting that the galaxies had much more mass than could be seen and that the mass stretched in a halo throughout a galaxy and far beyond its visible edge. Then, in 1974, theorists Jeremiah Ostriker, James Peebles, and Amos Yahil, who were all at Princeton University, published a landmark paper that opened with these words: "There are reasons, increasing in number and quality, to believe that the masses of ordinary galaxies may have been underestimated by a factor of 10 or more." The paper, along with Rubin's observations, became instrumental in turning opinion around and remains one of the most cited in the history of astronomy and cosmology. Years later, Rubin wrote: "By 1982, after a decade of initial disquiet, most astronomers reluctantly accepted the conclusion that a galaxy consists of much more than the luminous stars, gas, and dust that can be observed at various wavelengths." And while the mysterious matter outweighed luminous matter in galaxies by a factor of 10, on the scale of clusters of galaxies and superclusters dark matter seemed to dominate even more. The universe's unseen mass began to weigh heavily on astronomers' minds.

• • •

Bernard Sadoulet remembers the disquiet over dark matter continuing into the early 1990s, when he was the director of the Center for Particle Astrophysics in Berkeley, California. At talk after talk he gave in physics departments, he encountered skepticism over this potentially new form of matter. But even the sternest skeptics came around when it became clear that three independent lines of evidence all pointed in the same direction.

The first was increasing confirmation of Zwicky's work: The velocities of galaxies in a cluster suggested that much more mass was needed to keep them in its so-called gravitational potential well. A massive object creates a dent in the fabric of spacetime—much as a heavy ball placed on a rubber sheet dents the sheet. According to Einstein's general relativity, this dent—the gravitational potential well—is a measure of the object's gravity. The greater the mass, the bigger the dent and the deeper the well. Had the gravitational potential well of these clusters been shallower, the fast-moving galaxies would have long since careened off into space. Something was creating a well deep enough to keep the galaxies in it.

On the heels of these measurements came new data. NASA's Einstein Observatory, a space-based X-ray telescope that flew from November 1978 to April 1981, had measured the temperature of gas falling into such gravitational potential wells. Again, the deeper the well, the more the gas should heat up. The temperature of the infalling gas clearly showed the gravity of clusters to be much greater than could be explained from luminous matter alone.

The third piece of evidence came from gravitational-lensing experiments. Clusters dent spacetime in such a way that light coming from a remote galaxy behind a cluster will bend toward us. It's as if a gigantic lens stands between us and the faraway galaxy. The amount of lensing is a precise indication of the mass of the cluster: The greater its mass, the bigger the lens, and the more the light bends. Such lensing experiments showed that the mass of clusters was greater than what we saw out there. "These three pieces of evidence that have very different physics were all pointing to a similar depth of gravitational potential wells, and really convinced the community that this [depth] was not just artificial," Sadoulet told me in a recent phone conversation. "It was much more than that."

While observations kept indicating that dark matter existed, there were also tantalizing hints from cosmology and particle physics. One of the first clues as to the nature of dark matter came from the theory of big-bang nucleosynthesis. The theory lays down the precise sequence of nuclear reactions that produced the nuclei of such light elements as deuterium and helium in the minutes following the big bang. In the 1940s, soon after Edwin Hubble's pioneering work and Einstein's theories showed that the universe had a hot beginning, Gamow and his student Ralph Alpher laid the foundation for the theory of big-bang nucleosynthesis. When it came time to submit what was a rather speculative paper for publication, Gamow insisted on adding the name of his friend Hans Bethe, a scientist already famous for his work on nuclear fusion. Gamow was being a jokester, for the authorial last names would now constitute a play on the first three letters of the Greek alphabet, α, β, and γ. The young Alpher resisted the idea at first, concerned that he would be trumped by two famous men for what was essentially his work, but finally relented. The paper, titled "The Origin of Chemical Elements," was published on April 1, 1948. It was an instant hit, but despite its success, Alpher still had to defend the idea for his Ph.D. thesis.

To Alpher's horror, Gamow wasn't done with his pranks. He invited Bethe to attend the thesis defense with this rather cheeky note: "Dear Hans! This is a frame-up. We got your name on the element producing paper, and now the University wants you to take part in Alpher's examination since you will be unable to talk against it! Ain't it smart!" Though Alpher defended his thesis successfully, he never, as he had feared, quite got the recognition he deserved. But the theory of big-bang nucleosynthesis (BBN) was on its way.

BBN put an upper limit on the amount of protons and neutrons (collectively called baryons) — particles that make up the bulk of normal matter — which should have been produced after the big bang. But astronomical observations were clearly showing that there was much more matter out there. What was it, then, if it wasn't baryonic?

The flippant answer was that it was nonbaryonic—matter not made up of protons and neutrons (or their constituents, quarks). Still, what exactly *was* this matter? As it happened, another theoretical breakthrough, having to do with the standard model of particle physics, was also leading physicists toward a nonbaryonic candidate.

Developed in the 1960s and 1970s, the standard model beautifully explains all the observed particles that make up normal matter: protons, neutrons, and their quark components, along with the much lighter electrons—all of which are collectively called fermions. The model also explains the particles called bosons, which are the carriers of the fundamental forces (except gravity, which the standard model does not address). But there are a few niggling inconsistencies. As we will see in greater detail in chapter 9, elementary particles like the electron and the bosons get their masses from their interactions with an as-yet-undiscovered particle called the Higgs boson. But in the standard model, the mass of the Higgs boson is unstable—it diverges toward extremely large values, orders of magnitude more than what experiments tell us it should be. Physicists soon figured out that the mass of the Higgs can be made stable, in theory, if a whole new set of hypothetical particles was introduced. This new theory, which came to be called supersymmetry, added a fermionic partner particle to every boson and a bosonic partner to every fermion. These supersymmetric particles would have been produced in the early universe, but they would all have decayed—except for one. This hypothetical particle, called the neutralino, is neutral, stable, and very well suited to be the mysterious dark matter. It is massive enough, interacts weakly with matter, and is nonbaryonic. Also, it would have had the necessary "relic density" in the early universe to produce the amount of dark matter that astronomers estimate is out there today. What was most satisfying to physicists was that the neutralino popped out of theoretical considerations that had nothing to do with dark matter. Moreover, it emerged as a likely candidate well before cosmologists had concluded that dark matter had to be nonbaryonic.

Thus, two lines of reasoning, one from cosmology and the other from particle physics, both suggested that dark matter was made of slow-moving, weakly interacting, massive particles, which quickly came to be known as WIMPs. "These two rationales, coming from two different physics, were pointing to the same missing piece of the puzzle," Sadoulet said. "It may be a coincidence, but it was enough for me to spend twenty-three years of my career on."

A WIMP is a form of cold dark matter, to distinguish it from other kinds of dark matter. In some theories, dark matter is made of particles moving almost at the speed of light, which makes them highly energetic, or "hot." Cosmologists failed to explain the observed large-scale structure of clusters and superclusters of galaxies by using hot dark matter, causing the idea to fall by the wayside. However, they did so very successfully by using cold dark matter. These days, the terms "dark matter" and "cold dark matter" are used interchangeably (as they will be in this book, unless explicitly stated otherwise).

Convinced of the existence of cold dark matter, Sadoulet teamed up with physicist Blas Cabrera of Stanford University and set up the Cryogenic Dark Matter Search experiment that would eventually lead them to the Soudan Mine. They began to look for WIMPs. A neutralino is a WIMP. But even though WIMPs were the prime suspects, they were exotic enough to prompt astronomers to look for other possibilities. They wondered, for instance, whether dark matter wasn't simply normal matter that was somehow emitting little or no light—and this idea produced another locker-room acronym: MACHOs (for MAssive Compact Halo Objects). Among the possible MACHO candidates were brown dwarfs, which are substellar objects that don't have enough mass to sustain hydrogen fusion and thus are extremely faint. Other candidates included neutron stars, the immensely heavy and dense objects left over after stars have exploded and died. Researchers at Berkeley conducted a thorough search for MACHOs in the far reaches of our galaxy. The experiment was based on the idea that a MACHO warps spacetime and acts as a gravitational lens. If one comes between Earth and a star, it focuses the starlight toward us, causing the star to brighten momentarily. But the search came up short. There are just not enough MACHOs to explain dark matter. By the late 1990s, these objects fell out of favor, and WIMPs, still undiscovered, ruled the theoretical roost.

Richard Gaitskell is one of the most colorful members of the CDMS team. He's a tall Englishman of ample proportions, with a ready, generous laugh, a quick wit, and a decidedly odd taste in socks. I met him one October evening at his office at Brown University in Providence, Rhode Island. He was neatly turned out in shirt, tie, sweater, corduroy trousers, and shoes, all in various shades of brown. The finishing touch was a pair of pink socks. Gaitskell, the head of the Particle Astrophysics Group at Brown, has an experimentalist's chutzpah and a theorist's love for ideas. As he was growing up in England, his first serious experiment was on religion. "At the age of thirteen or fourteen, I did due diligence for about a year," he told me. "I hung around with a lot of people who had clearly spent a lot more time than I studying religion—for the most part, Christianity, given the school I was in." But after "a year or two of cross-examination," Gaitskell decided that whatever they were getting out of religion wasn't something he needed.

Gaitskell turned his adolescent attention to numbers, specifically stock market indices. He grew to love them so much that after college he went straight to being an investment banker. But even the financial high life of the 1980s in London didn't quite do it for him, and he turned back to Oxford. "As I said to the fellowship selection committee at Oxford, 'Greed became boring,'" he recalled. Gaitskell got the fellowship, and for the first time got pulled into science. "One of the beauties of the Oxford-Cambridge system—and this is the thing that got me addicted—was that I could spend all week in the lab. I would get out of bed, grab a croissant and coffee on the way, disappear into the lab, and not reappear until the sun was down, eleven o'clock at night. I think I'm clearly obsessive. That helps with science."

Oxford encouraged this kind of single-minded focus. By opting to live in college housing, he eschewed landlords or landladies, neighbors, or having to feed himself. The food was catered by the college. "I was a bit like the Queen," he said. "I never used to carry any money. The things I was living on back then, when I had a rather better physique, were cigarettes and booze. I could buy both on credit at the college. At some stage, the money would disappear from your bank account."

With a doctorate in physics under his belt, Gaitskell turned up at Berkeley in 1995 to work with Sadoulet, Cabrera, and the CDMS team. The researchers had decided to tackle the problem of dark matter by trying to find a WIMP in the lab. Easier said than done. By their very definition, WIMPs do not interact with matter electromagnetically, so they will pass right through stuff. "I once worked out that if you piled lead from the Earth all the way to the sun, and if you fire a WIMP at the stack of lead, it still has a fifty-fifty chance of coming out of the stack without interacting," Gaitskell said.

Not the kind of odds a sensible person would accept if asked to snag a WIMP. But experimentalists are a doggedly optimistic breed. Theory tells them that our galaxy should have an extensive halo of WIMPs, so WIMPs should be passing through Earth all the time. In fact, about 4 billion WIMPs, each with a mass of about 100 gigaelectron volts (GeV), should be passing through the CDMS detectors every second. All one has to do is wait for one of these particles to hit a detector head-on. "What we are looking for is the very occasional interaction of these WIMPs with a nucleus. And when it does collide, it will transfer most of its energy," said Gaitskell. It's not unlike a cue ball hitting another ball on a billiard table: the cue ball comes to a dead stop, and the second ball rockets away.

It was to test this idea that Sadoulet and Cabrera designed the first CDMS experiment. Conceptually, their detector was simple. A germanium crystal was cooled down to less than a few tenths of a kelvin, or nearly $-273°C$. A WIMP smashing into a nucleus in the crystal would raise the temperature of the crystal by a few microkelvin, a difference that could be detected. But there was one big problem — there are many different particles that can create the same signal. The whole exercise, in a sense, came down to ruling those out. Every possible source of a false signal had to be either prevented from hitting the detector or characterized and understood so thoroughly that it could never be mistaken for dark matter.

One of the biggest sources of false signals, or noise, is ambient radioactivity. Even humans are radioactive enough to be a problem. So the germanium detectors had to be shielded inside layers of copper and lead. But lead can contain lead-210, a radioactive isotope with a half-life of twenty-two years (that is, after twenty-two years, half of a given amount of it will have radioactively decayed). So the team had to find old lead, in which most if not all of the lead-210 had decayed already. An Italian colleague mentioned that he had been using lead taken from two-thousand-year-old Roman ships that had sunk off the Italian coast. The CDMS team located a company that was selling lead salvaged from a ship that had sunk off the coast of France in the eighteenth century. Unaware that they were doing anything illegal, the researchers bought the lead. The company, however, got in trouble with French customs for selling archaeological material.

Illegal or not, the lead worked: The detectors were now shielded by high-purity material, with the radioactivity more or less under control. "At that stage, you are basically dominated by some idiot who licks his finger and does this," said Gaitskell, mimicking someone touching a detector with a wet finger. "The radioactivity in that fingerprint is going to show up." That's how sensitive their detectors were.

Besides ambient radioactivity, there were muons to contend with. These are subatomic particles, created when cosmic rays—particles such as electrons, protons, and helium nuclei from outer space—hit Earth's atmosphere. Such muons could smash into nearby rock and spew out neutrons, which could then hit the detectors and easily mimic a dark-matter particle. So while the CDMS team had engineered beautiful germanium crystals and understood a lot about suppressing noise from background radioactivity, cosmic rays were still a huge problem. Their experiment, which was in an underground lab about 30 feet beneath the Stanford campus, was still too near the Earth's surface. That's when their attention turned to the Soudan Mine.

• • •

While the blizzard raged up above, inside the mine I was about to come face-to-face with the team's second-generation experiment— CDMS II. Curiously, CDMS II is doing its damnedest not to detect anything—not background radiation, not neutrons kicked off by cosmic-ray muons hitting the rock. Essentially, nothing that can be construed as a WIMP. The idea was to build a detector assembly so quiet that nothing registered on it. Only after you have built one that can go for months without seeing anything can you be sure that when you do see something, it is most likely a WIMP. But how do you build an experiment that detects nothing?

To learn about the practical aspects of such an enterprise, I met the CDMS technician, Jim Beaty, a tall, strapping Minnesotan with an accent straight out of the Coen brothers' *Fargo*. Beaty had been with the underground facility for about nineteen years. Working in mines ran in the family. Beaty's father, grandfather, and an uncle had all been miners. But Beaty's was a different sort of mining job.

"My kids ask, 'What do you do?'" he told me. "How do you explain something like this? Being a child-minder for a whole bunch of physicists—how do you explain it? I can't," he said, only half facetiously. As a technician for an underground physics laboratory half a mile below the surface and miles from the nearest city, he found himself having to constantly innovate to satisfy the technological whims and fancies of experimental physicists. "But it's kind of nice to get up in the morning and actually look forward to going."

"Into a hole?" I asked.

"Yeah, into a hole."

Beaty had seen the lab take shape from the early days, when scientists brought an unusual experiment to the mine. There are theories that try to explain why our universe is made of matter rather than antimatter by suggesting that nature might have favored the production of matter over antimatter by just a smidgen. All the antimatter would have annihilated with matter, and the leftover matter gave us our universe. The same theories also predict that protons should decay. So this experiment, called Soudan II, was designed to look for the decay of protons by monitoring nearly 1,000 tons of iron, containing about 6×10^{32} protons.

Beaty still remembers the nerve-wracking challenge of lowering each of the 224 modules that made up the megaton of iron down the mine shaft. Trucks would bring the modules, each weighing 4.3 tons and designed to just fit into the shaft cage, to the mine's headframe. Beaty's team, using a salvaged crane and working with clearances of an inch or two, loaded them into the cage (knowing that they cost $50,000 apiece didn't help). "Each night we would do four, and after doing four, I'd be tense and sweat-soaked," Beaty recalled.

On top of this, he had to contend with suspicious locals. Some of the Soudan II modules had tiny stickers of mildly radioactive iron-55 needed for calibration, so the trucks had to be labeled as carrying radioactive cargo. And because the Minnesota Department of Natural Resources was conducting tours of the mine during the day, Beaty's team had to work at night, unloading the trucks with just two lamps mounted on the crane for lighting—they couldn't afford big spotlights. The veneer of secrecy and the warnings about radioactivity were too much for some of the natives. "Occasionally, I'd be up there at night, running the crane, and I would see a flash out in the woods," said Beaty. "Somebody was taking pictures. It was a rumor for years that we were bringing nuclear waste underground."

Everything that makes up the Soudan physics lab and its experiments has to come down the shaft, in cage-size pieces, even if it means dismantling some of the equipment. Underground, I saw entire offices, air conditioning, I-beams with overhead cranes, a forklift truck, and even a Ping-Pong table, all of which had presumably undergone subterranean assembly. During the excavation of the caverns, Beaty told me, the crew brought down a full-size front-end-loader tractor in pieces.

Amid all this massive apparatus, CDMS II stands apart. At its heart are the thirty detectors, like exquisite and fragile works of art. Each detector—a centimeter thick, about the diameter of a hockey puck, and made of either 250 grams of germanium or 100 grams of silicon—is a painstakingly grown, ultra-pure crystal. Photolithographic techniques are used to pattern thin superconducting films of tungsten and micron-thin "fins" of aluminum onto the surface of the crystals. The fins can break if the detectors are so much as bumped, so the detectors were shipped from California to Soudan by a trucking company that specializes in handling expensive art. Each climate-controlled truck carried one set of detectors, worth a quarter of a million dollars. Usually, two drivers alternated, driving the truck nonstop to minimize the amount of time the detectors spent above-ground and thus exposed to cosmic rays. As always, Beaty was on hand to collect the precious cargo. When he first opened the doors of the trailer, he was struck by how small the detectors ("our little thing") looked strapped down in the middle of the "gorgeous oak-lined huge forty-four-foot" interior. Even my host, Mike Dragowsky, a man who speaks in measured tones, had been moved by the pieces of precision engineering. "I think the detectors are the neatest things that I have ever encountered," he said.

CDMS II is looking for the subtle vibrations of a crystal's lattice that would occur if a dark-matter particle smashed into it. But these vibrations would be so feeble that the normal random thermal motions of molecules in the crystal could easily swamp them. So the detectors, vastly improved since the first CDMS runs at Stanford, have to be cooled down to about 40 millikelvin (mK), a touch above absolute zero. (It's a cold that defies imagination; the detectors are nearly 273°C lower than the temperature at which water freezes.) The extraordinary cold stills the crystal's thermal jiggling, so that if a WIMP smashes into a germanium or silicon nucleus, the vibrations from the collision will be picked up by the thousands of superconducting sensors on the crystal's surface. This is not unlike chucking a stone into a lake to see the resultant ripples. On a choppy lake, the ripples are barely noticeable, but on a still day, when the lake is as smooth as a sheet, even the tiniest of ripples are obvious. As noted, vibrations in the crystals can also be caused by more mundane particles; understanding and excluding every particle that can cause the same signal as a WIMP is at the heart of building an experiment that detects nothing.

To see how this exclusion was being done, I followed Dragowsky into the radio frequency (RF) room, a cage that isolates its inside from the outside, electromagnetically speaking (embassies around the world use RF rooms to prevent electronic snooping). Inside the cage is a clean room, protected by an air lock. To avoid contaminating the equipment, we wore powder-blue slip-ons over our shoes, stepped on sticky pads to get rid of any dust, and donned gloves and hairnets—all this before entering the anteroom. Inside the anteroom, we slipped into clean coveralls and put yet another cover on our shoes. My camera and tape recorder had to be wiped with alcohol to avoid bringing any radioactive mine dust near the experiment.

Inside the clean room, there was a hush, broken only by the *swoosh-swoosh* of a cryocooler, which was circulating high-pressure helium gas for one of the cooling stages. Everywhere I looked, I saw a tangle of gleaming tubes and wires, all of which seemed to disappear into a large container a couple of meters across. Somewhere inside was what the researchers fondly called the "icebox," a rather understated term for what is surely one of the coldest places on Earth.

Multiple stages of cooling, using cryogenic liquids, keep the innards of the icebox at about 40 mK. To achieve this temperature, each stage is built like a thermos, insulating what's inside from what's outside. Liquid nitrogen cools everything down to about 77 K, then liquid helium brings the temperature down to about 4 K. The very last stage of cooling uses a supercooled mixture of helium-3 and helium-4. Deep inside the icebox are the germanium and silicon detectors, five towers encased inside pure copper tubes, each tower holding six crystals. The copper is the link between the icebox and the detectors, keeping the crystals cold.

For CDMS II, the purest possible copper had been mined, processed into sheets, and sent to Soudan. It had to be transported by ship and road, not air, to keep exposures to cosmic rays to a minimum. Cosmic rays get more numerous as you go higher up in the atmosphere, and they can hit nuclei within materials, spewing out neutrons, which can then interact with other nuclei and lead to the formation of radioactive isotopes. Radioactivity, as I was repeatedly reminded, is a huge problem for CDMS. Also, the copper has to be handled carefully. Beaty, who joined us in the RF room, said, "I was in here one day cleaning some of the copper, and we were told, 'Don't

rub too hard, because you don't want to harden the copper and destroy its ability to do its job.'" After the admonishing, he said, "we
were sitting in here, going like this," and he mimicked the motions
of someone carefully wiping a fragile and expensive antique, "terri-
fied that we were going to harden that copper."

Although the level of cosmic rays inside the mine is already considerably less than it was in the underground Stanford lab (the immense amount of rock overhead is a formidable barrier), some cosmic-ray muons still make it through. These can hit the material
surrounding the detectors, ejecting low-energy neutrons, which can
then strike a detector and mimic dark-matter particles. The only way
to avoid this sort of false signal is to track these rogue muons. This
is done by panels of so-called plastic scintillators, which surround
the icebox and give off light when a muon passes through them.
Any particle colliding with the detectors that coincides with a muon
streaking through the panels is discarded. It's deemed an imposter,
not a WIMP.

But the muons can also strike the rock outside the plastic-scin-
tillator panels and generate low-energy neutrons. So just inside the
panels is a layer of polyethylene about half a meter thick. Any neutrons headed toward the crystals will be scattered by the hydrogen
molecules in the polyethylene, so that by the time they emerge they
won't have enough energy to register on the detectors. With muons
and neutrons more or less taken care of, the next enemy is ambient radioactivity from the radon in the mine and from the surrounding 2.7-billion-year-old Ely Greenstone. To fend off radioactivity, a
17.5-cm-thick layer of lead is placed inside the polyethylene. There
is another 5-cm-thick layer of old lead, salvaged from sunken ships,
to shield the detectors from any radioactivity from the newer lead.
Inside this second layer is yet another thin layer of polyethylene, to
catch any neutrons that might be produced in the ancient lead itself.

But even the most diligently designed shield can still let through
particles that are not WIMPs. If a particle does enter the detector
assembly, the CDMS team has to figure out what it is. Fortunately,
there are ways to discriminate between, say, photons or electrons

and a WIMP. A photon or electron will bounce off the cloud of electrons in germanium or silicon atoms. This is called an electron-recoil event and can be distinguished from the nuclear-recoil event expected of dark matter. Electron-recoil events, too, will make the lattice vibrate, but they will also produce ionization within the crystal. The researchers realized that they could measure this ionization and use it to distinguish an electron-recoil event from a nuclear-recoil event. Problem solved. But what about the odd neutron that makes it through all the shielding, soars past the electron cloud, hits a nucleus, and mimics a WIMP?

That's where the germanium and silicon come in. WIMPs are by definition weakly interacting. In other words, when they hit a nucleus, they interact via the weak nuclear force, and the interaction depends on the size of the nucleus. Neutrons, on the other hand, interact with a nucleus via the strong nuclear force, another fundamental force of nature, and the interaction does not depend much on the size of the nucleus. Given that germanium nuclei are much bigger than silicon nuclei, a WIMP is more likely to interact with germanium than with silicon. Neutrons will interact similarly with both crystals. So over the course of the experimental run, if there is an equal number of collisions with nuclei in both silicon and germanium, then they are unlikely to be WIMPs. But if there are more events in germanium than in silicon, the excess nuclear recoils are most likely being caused by WIMPs.

What the CDMS team is trying to build is an apparatus in which every possible background particle is screened out. After screening and subtracting, there should theoretically be nothing left—nothing, that is, except for dark-matter particles. The sensitivity of the experiment is measured in terms of how many days the detectors function with no signals from background particles. Things at CDMS II are looking good for a 1,700-kilogram-day run, which is the equivalent of a 1,700-kilogram detector running for a day, or a 1-kilogram detector running for 1,700 days, or any combination in between. At the end of the run, if the team sees a few events, they can confidently say that they have seen WIMPs. At least that's the hope.

So far, there have been no signs of a WIMP (the Soudan II experiment, which ran from April 1989 to June 2001 and looked for proton decay, also failed to find anything). In March 2008, the collaboration reported the first results from their full complement of thirty detectors. WIMPs had eluded a 121.3-kilogram-day run of the experiment. Despite this, the scientists are compelled to continue. Dragowsky typifies the experimentalist's mindset. "Ultimately, you have to take some pleasure from making the system that supports the detectors," he said. "You have to be overwhelmingly driven by the idea that you want to make this measurement. It allows you to push through the frustrations that are associated with making a complicated, delicate apparatus come to life."

If dark-matter particles are indeed streaming right through us unhindered, then there's a chance, a slim chance, that the delicate crystals of germanium and silicon deep within the icebox will see some. It might take more than a decade to find just a few. "I think it'll be interesting whether it is a positive answer or negative answer," said Dragowsky. "If we don't find evidence in the dark-matter detectors that are operated over the next ten years, that's interesting." Because if there's one thing he's convinced of, he said, it's the existence of dark matter in the universe.

If direct dark-matter detectors like CDMS II and others in various underground laboratories around the world never see any WIMPs, the experiments will put theoreticians on notice. They will have to work harder to explain just what kind of dark matter is out there—and if, indeed, dark matter is made of WIMPs. As Gaitskell reminded me, the question about whether dark matter is made of particles that at least feebly interact with ordinary matter is still an open one, despite all the effort that went into crafting these detectors. "Does it make sense to be looking for them with direct detection?" he mused. "That's a total crapshoot. Thirty years from now, when one picks up a book on cosmology, the whole WIMP thing could literally be a footnote."

But footnotes are not what these experimentalists are banking on. The search for dark matter is proceeding on multiple fronts. The world's biggest particle accelerator, the Large Hadron Collider (LHC), is humming near Geneva. If physicists are right about the universe being supersymmetric, then the LHC has a very good chance of creating the supersymmetric particle—the neutralino—that could be dark matter. "Most people in the direct-detection field [like CDMS] would find it highly amusing to see dark matter directly, and actually beat the multibillion-dollar LHC to the discovery," said Gaitskell.

I would eventually make it to the LHC to see how confident those scientists were about scooping experiments like CDMS II. But first I headed to Siberia, where physicists are using the world's biggest freshwater lake to look for subatomic particles called neutrinos from the center of our galaxy. This would be an indirect confirmation of dark matter, for if WIMPs exist they should be accumulating in great numbers at the heart of the Milky Way, because of the gravity of the supermassive black hole there. The mutual annihilation of densely packed dark-matter particles should be spewing out neutrinos.

Decades or hundreds of years hence, when dark matter is no longer mysterious and underground experiments have eased into history, their success or failure notwithstanding, some intrepid explorers might enter a long-forgotten mine like Soudan. They would find dead, desiccated bats hanging on the walls of the cavern and a massive mural—a modern-day cave painting—depicting the progress in particle physics. Somewhere on the floor they might find blueprints of the experiment that detected nothing—all evidence of our baby steps up the steep staircase of knowledge.

3 · LITTLE
NEUTRAL ONES

ABOUT 25 MILLION YEARS AGO, the earth parted in the southeast corner of Siberia. Since then, countless rivers have converged on the gaping continental rift, creating the massive body of water known as Lake Baikal. Surrounded by mountains, this inland sea has forever been isolated from other lakes and oceans, leading to the evolution of unusual flora and fauna, three-quarters of which are not found elsewhere on Earth. Russians regard it as their own Galápagos. The lake contains 20 percent of the world's unfrozen fresh water, and just a little less during the severe Siberian winter, when — despite its enormous size and depth — Baikal freezes over.

On one such winter's day, two days into my trip to the region, I found myself on the lake near the town of Listvyanka, which is nestled in a crook in the shoreline. I was in an old van that was trying to head west — not along a coastal road, for there was none, but over the ice. The path, however, was blocked by a ridge. It looked like a tectonic fault; two sections of the lake's solid surface had slammed together and splintered, throwing up jagged chunks of ice. The driver, a Russian with a weather-beaten face, peered from underneath his peaked cap, looking for a break in the ridge. When he spied a few feet of smooth ice, he got out and prodded it with a metal rod, only to shake his head as it crumbled — not thick enough to support the van. We kept driving south, farther and farther from shore, in what I was convinced was the wrong direction. The van shuddered and lurched along, its tires crunching on patches of fresh snow and occasionally slithering on ice. The ridge continued as far as the eye could see. Suddenly, we stopped.

In front of us was a dangerous-looking expanse littered with enormous pieces of ice that rose from the lake's frozen surface like giant shards of broken glass. The driver seemed to be contemplating going around them to look for thick, solid ice that would let us continue to our destination, an underwater observatory operating in one of the deepest parts of the lake. But if he did that, we'd get even farther from the shore, and it would take just one punctured tire to strand us. The sun was little more than an hour from setting, and the temperature was falling. I couldn't ask him if he had a radio or a phone to call for help, since he did not speak a word of English and the only Russian phrase I knew was *Do svidania*. The last thing I wanted to say to him at this point was "Goodbye."

Thankfully, he decided to turn around. We drove along until we came upon vehicle tracks that went over some ice covering the ridge. The driver swung the van westward and cleared the ridge, and soon we were racing across the lake at speeds that turned every frozen lump into a speed bump. The van's front rose and fell sickeningly, rattling the tools strewn around on the front seat. I worried that the ice would give way and we'd plunge into the frigid waters below. But it remained solid, and the van, despite its appearance, was in fine mechanical fettle, its shock absorbers holding firm. In the distance, I spied a dark spot on the otherwise white expanse. As we approached, the spot grew to its full size, revealing itself as a 3-foot-high Christmas tree. We still had 35 kilometers to cover, and the sun would soon disappear below the icy horizon. But now that we had found the Christmas tree, I knew we were fine.

I had first seen the tree two days earlier, with Nikolai (Kolja) Budnev, a physicist from Irkutsk State University, and Bertram Heinze, a German geologist. We were headed to the site of the Lake Baikal neutrino observatory, which lay deep beneath the ice. We had just driven onto the lake from the shore near Listvyanka when Heinze asked, "When does the ice start breaking?"

"Sometime in early March," Budnev answered.

My heart skipped a beat. It was late March, and we were on the ice in an old olive-green military jeep. "Sorry, sometime in early April," Budnev corrected himself. *Phew*.

Earlier that morning, Budnev had picked me up at Irkutsk's dilapidated airport. I had flown in from Moscow, and the contrast between the capital's glitzy new international terminal at Domodedovo airport and the one in Irkutsk was stark. The dusty-yellow tarmac bus that ferried us from the plane to the Irkutsk terminal was so overloaded that its rear scraped the ground, and the terminal's concrete structure was falling apart. (The Irkutsk airport was in the news three months later; the very flight I had flown in on crashed after landing, killing 125 people and injuring 78 others.) In the past, when Russia's influence spread beyond Siberia into Alaska and even California, Irkutsk had been a prosperous city. On this cold, gray morning in 2006, the city center looked rundown, its decaying wooden buildings only hinting at that impressive past.

We stopped to collect Heinze from his hotel. A tall man, he had to scrunch to fit into the jeep's front seat. Soon we were speeding away from Irkutsk toward Listvyanka, on a wide road with forests on either side, the speckled white of the winter-bare birches broken occasionally by evergreen pines. We soon overtook a van—the same van that would be a source of angst two days later—taking other researchers to the observatory. In the van's rooftop rack were planks of wood. "See those planks?" said Budnev. "We use it to cross cracks in the ice."

Before I could voice my anxiety, Budnev went on to explain that if the cracks were less than about a foot and a half wide, the planks were not used. The driver could speed up and clear such cracks, he said, embellishing the explanation with sound effects—the *swoosh* of a vehicle taking off and landing with a thud on the other side.

Could the ice handle the weight of the jeep? I wondered aloud. Budnev assured us that the ice could support a tank. In fact, it could support a train. Well, almost. Before the Trans-Siberian Railway was completed, the tracks stopped on the western shore of Lake Baikal, then continued from the eastern shore. The entire train—carriages, engine, cargo, and passengers—was ferried by specially built icebreakers across the lake. Once, in the winter of 1904, during the Russo-Japanese War, the ice had held firm and reinforcements were urgently needed on the other side, so the Russians laid railroad tracks on the frozen lake. An engine was sent to test the ice. It never made it across and still lies somewhere on the bottom. If I had known this at the time, Budnev's words would have seemed less reassuring.

However, it is the ice that makes possible an extraordinary enterprise. For more than two decades now, Russian and German physicists have camped on the frozen surface of Lake Baikal from February to April, installing and maintaining instruments to search for the "little neutral ones"—elusive subatomic particles called neutrinos. Artificial eyes deep below the surface of the lake look for tiny flashes of blue light, caused by an unlikely collision between a neutrino and a molecule of water. I was told that human eyes could see these flashes, too, if our eyes were the size of giant watermelons. Indeed, each artificial eye is more than a foot in diameter and the Baikal neutrino telescope, the first such instrument in the world, has 228 eyes patiently watching for these messengers from outer space.

The telescope, which is a few kilometers offshore, operates underwater all year round. Cables run from it to a shore station, where the data is collected and analyzed. It's a project on a shoestring budget. Without the luxury of expensive ships and remote-controlled submersibles to allow them to work on the telescope, the scientists wait for the winter ice to provide a stable platform for their cranes and winches. Each year, they set up an ice camp, haul the telescope up onto the ice from a depth of 1.1 kilometers, carry out routine maintenance, and lower it back into the water. And each year they race against time to complete their work, towing their equipment back to the shore before the sprigs of spring begin to brush away the Siberian winter and the lake's frozen surface starts to crack.

What is it about the neutrino that makes scientists brave such conditions? Neutrinos go right through matter, traveling unscathed from the time they were created — some of them right after the big bang — and carrying information in a way no other particle can. The universe is optically opaque at extremely high energies, for ultra-energetic photons are absorbed by the matter and radiation that lie between their source and Earth. However, neutrinos, which are produced by the same astrophysical processes that generate high-energy photons, barely interact with anything along the way. They can escape even from within astrophysical objects. For instance, neutrinos stream out from the center of the sun as soon as they are produced, whereas a photon needs thousands of years to work its way out from the solar heart. Neutrinos represent a unique window into an otherwise invisible universe.

While these particles are essential to astrophysics, they have something important to contribute to cosmology as well: They could unveil dense pockets of dark matter in our galaxy. Theory suggests that over time the gravity wells created by Earth and the sun would have sucked in an enormous number of dark-matter particles. There is an even bigger gravity well nearby (relatively speaking), created by the supermassive black hole at the center of the Milky Way. Dark-matter particles should be clustered there, too. Wherever they gather in great concentrations, these particles should collide with one another and annihilate, spewing out, among other things, neutrinos. It's as if a giant particle accelerator at our galaxy's center were smashing dark-matter particles together, generating neutrinos and beaming them outward, some toward us. Detecting neutrinos coming from the center of our galaxy would be akin to detecting dark matter, albeit indirectly.

Fortunately for physicists, neutrinos are known particles, unlike those that make up dark matter. But neutrinos come in a range of energies, and those expected to be produced by the collision of dark-matter particles would have about the same energy as those produced nearer home, in the atmosphere. When cosmic rays strike the upper atmosphere, they generate a secondary shower of particles, some of which are neutrinos. These so-called atmospheric neutrinos make up by far the greatest proportion of neutrinos arriving on Earth. They have to be studied in great detail, and their properties and statistical distribution across the sky diligently mapped, because any neutrinos produced by dark matter at the galactic center will appear as a mere blip against this background. Without a clear understanding of the background, the dark-matter signal will be missed.

There is yet another way in which neutrinos might illuminate the search for dark matter. It has to do with the standard model of particle physics. As noted, nothing in the standard model explains dark matter; it's clear that only new physics can do so. Just what this new physics is depends on discoveries that reveal other inadequacies of the standard model. Neutrinos can help here, because they, too, have properties that cannot be explained by the standard model. That neutrinos play such a key role in advancing physics would have surprised physicists of two generations ago. For them, the neutrino was a figment of imagination, a theoretical necessity but one that seemed impossible to detect because of its ethereal nature—a "ghost" of a particle.

The story of the neutrino begins in the late 1920s. Physicists had been puzzling over something called radioactive beta decay, in which an atom changes from, say, carbon-14 to nitrogen-14. Carbon-14 has eight neutrons and six protons. During beta decay, one of these neutrons decays into a proton and emits an electron. The new nucleus, now with seven protons and seven neutrons, is transformed into nitrogen-14. But during this process, some energy seemed to go missing, violating the law of conservation of energy. It was the Austrian-born physicist and Nobel laureate Wolfgang Pauli who finally figured out what was going on.

On December 4, 1930, Pauli penned an intriguing letter to colleagues who had gathered for a conference on radioactivity in Tübingen, Germany. Pauli began the letter "Dear radioactive ladies and gentlemen," and wrote about how he had hit upon a "desperate remedy" to rescue the law of conservation of energy. He theorized that the beta-decay process must emit an as-yet-undiscovered neutral particle. Though Pauli called this hypothetical particle a neutron, he was not talking about the neutron we know today as one of the constituents of atomic nuclei (this particle would be discovered in 1932). Pauli was aware that his idea was highly speculative. "I admit that my remedy may seem only slightly probable, because if neutrons do exist they should certainly have been observed long ago," he wrote.

A couple of years later, the theorist Enrico Fermi jokingly renamed the particle a neutrino, Italian for "little neutral one," and the name stuck. But for decades, the neutrino remained a theoretical construct, a useful particle that helped physicists save their theories from embarrassment. No one had seen one. No one even knew how to find one. Theory suggested that some neutrinos, especially low-energy ones, could go through lead light-years thick, so it was no wonder that experimental physicists were pessimistic about finding these ghostly particles. Unless, of course, something on Earth was producing neutrinos in unimaginably large quantities.

And there was. By the 1950s, America's nuclear weapons program was well under way, and Frederick Reines, a researcher in his early thirties working at the Los Alamos Laboratory in New Mexico, realized that a nuclear bomb would be a significant source of neutrinos (anti-neutrinos, to be precise, but treating them as neutrinos — as we will do throughout the book — does not compromise the explanation). Reines and his colleague Clyde L. Cowan Jr. reasoned that a nuclear power plant would also be such a source. They calculated that a detector near a nuclear reactor would encounter nearly 10^{13} neutrinos per square centimeter per second. There was just one small problem. Neutrinos are electrically neutral, which is one of the reasons they pass so cleanly through matter. They could be seen only if they directly hit the nucleus of an atom. Reines and Cowan would have to look for the signature of such a collision. And they found it. On June 14, 1956, the two scientists announced that their "Project

Poltergeist" had tracked down this ghost of a particle. The duo sent a telegram to Pauli, telling him of their results. Pauli happened to be at a meeting in Europe when the telegram arrived. He told everyone at the meeting the news, and later he and his colleagues celebrated by drinking an entire case of champagne.

In 1995, Reines would receive the Nobel Prize for this momentous discovery. (Cowan had died two decades earlier.) Between 1930, when Pauli decided that neutrinos must exist, and the Nobel for Reines, the neutrino had only grown in importance. Physicists were increasingly persuaded that neutrinos were central to the very fabric of the universe. They seemed to be everywhere. Some neutrinos were being produced right here on Earth; others had been created mere seconds after the big bang. And processes that span every instant of time, from the big bang to this moment, have made or are making neutrinos, from the fusion reactions that power the sun to the stupendous cosmic explosions called supernovae that signify the death of massive stars. The biggest source of Earthbound neutrinos is the sun. In any given second, close to a trillion solar neutrinos go right through just the palm of your hand.

By the 1960s, physicists had started building neutrino detectors inside mines as a natural way to shield them from cosmic rays, which wreak havoc just as they do in dark-matter experiments. In 1968, Raymond Davis and his colleagues from Brookhaven National Laboratory completed one inside the Homestake Gold Mine in Lead, South Dakota. They used a tank containing 100,000 gallons of tetrachloroethylene, a common dry-cleaning agent. When a neutrino smashed into an atom of chlorine, the atom was transformed into one of radioactive argon. By counting the number of argon atoms that had been produced, the physicists could calculate the number of neutrinos that were coming from the sun (after accounting for the fact that only a tiny fraction were interacting with the chlorine). And they were surprised to find far fewer than expected. This came to be called the mystery of the missing solar neutrinos.

Meanwhile, neutrino detectors got bigger and better. One of them was the Kamiokande detector, situated inside an active zinc mine in the Japanese Alps. Another was the Irvine-Michigan-Brookhaven (IMB) experiment, sited deep within the Morton Salt Mine, near Cleveland, Ohio. Still another was the Baksan Neutrino Observatory deep underground in the Russian Caucasus. On February 23, 1987, all three detectors saw something unexpected: a burst of neutrinos from outside the solar system. These neutrinos had come from the supernova SN1987A, which had exploded in the Large Magellanic Cloud. To this day, these remain the only neutrinos from outer space — besides solar neutrinos — that have been seen on Earth. The experiments also confirmed that the mystery of the missing solar neutrinos was real.

It took two even bigger detectors — the Super-Kamiokande (Super-K), which began operating in 1996 inside the same mine as the Kamiokande detector, and the Sudbury Neutrino Observatory (SNO), built inside a nickel mine in Ontario, Canada — to solve the mystery. According to the standard model of particle physics, neutrinos come in three types: electron, tau, and muon. Between them, Super-K and SNO were sensitive to two of the three types of neutrinos, and their combined data were able to show that solar neutrinos were changing from one type to another as they sped toward Earth, a phenomenon known as neutrino oscillation. Even though the sun was producing the predicted amount of a given type, the neutrinos were morphing on their way to Earth, leading to the deficit in detectors that were designed to observe only one type, the electron neutrino. The Super-K confirmed oscillations in atmospheric neutrinos as well.

The observations of oscillating neutrinos posed a serious problem for the standard model. Neutrinos, according to the model, have no mass, but they can oscillate only if they do have mass. Hitoshi Murayama, a theoretical physicist at the University of California, Berkeley, was at the Super-K team's announcement in Japan in 1998. He wrote later, "It was a moving moment. Uncharacteristically for a physics conference, people gave the speaker a standing ovation. I stood up too. Having survived every experimental challenge since the late 1970s, the Standard Model had finally fallen. The results showed that at the very least the theory is incomplete."

The Japanese had perfected the method of using tens of thousands of tons of water in a tank lined with photomultiplier tubes (PMTs). The PMTs look for light emitted when a neutrino smashes into water. Normally, the neutrino will pass right through water without any interaction. But when one does occasionally hit a nucleus of hydrogen or oxygen, the collision can sometimes spit out another subatomic particle, a muon. The charged muon interacts with the water electromagnetically, and because it is moving faster than the speed of light in water, it leaves in its wake a cone of blue light. This is called a Cherenkov cone, after the Russian physicist who first described the phenomenon. It is analogous to the sonic boom caused by an aircraft traveling faster than the speed of sound.

Even as the Super-K and SNO experiments were being built, physicists were training their sights beyond the sun. But this brought new challenges. Most astrophysical neutrinos, including those produced by any dark matter at the galactic center, are higher in energy than solar neutrinos. While such neutrinos are easier to detect (because they produce brighter and longer streaks of light in water), their numbers fall dramatically with increasing energy. The only chance of seeing one is to monitor far greater volumes of water than is possible in underground detectors, which are limited in size by the mines that confine them. Where could one find such an abundance of clear water?

In the 1960s, the Russian physicist Moisey Alexandrovich Markov, a "poet" of astroparticle physics, had suggested using natural bodies of water as neutrino detectors. Instead of building tanks of water inside mines, why not just use lakes, or even oceans? Just submerge long strings of photomultiplier tubes into the water and watch for the Cherenkov light left behind by neutrino-generated muons. It was an enticing idea, but there were enormous practical difficulties. For one thing, without rock above to protect it, a detector would be exposed to cosmic rays that could swamp any signals from neutrinos.

When cosmic rays strike the upper atmosphere, they generate neutrinos, but they also generate muons in roughly the same numbers. Cosmic-ray muons and atmospheric neutrinos will both reach the water, but for every neutrino that interacts and creates a muon that lights up the water, a billion cosmic-ray muons will do the same. So if a neutrino detector is placed near the surface of a lake, it will be overwhelmed by the Cherenkov light from cosmic-ray muons striking the water and unable to distinguish any muon generated by a neutrino. More to the point, sunlight (which is not a problem inside mines) will blot out the Cherenkov light. The solution for both problems is to go deep, where the sun's rays cannot reach. Water also blocks many of the cosmic-ray muons, and about a thousand times fewer muons reach the bottom of a deep lake. But even that is too many.

Physicists realized that they could use Earth itself as a shield. While many muons can make it through a kilometer of water, a similar stretch of rock will stop them cold. So a neutrino detector can sit near the lakebed, positioned to look for muons created by neutrinos that come from below. None of the muons created by cosmic rays in the atmosphere on the other side of Earth can penetrate the planet. Neutrinos, however, zip right through, and occasionally one will hit a nucleus in the water or in the lakebed itself. Such a collision generates a muon, which then shoots up toward the surface. Catch an upward-moving muon and you have essentially detected a neutrino that came from the other side of Earth.

All that was needed was a suitable body of water. By the mid-1980s, the Russians realized that they had a massive tank (636 kilometers long and 80 kilometers at its widest) of pure water in their own backyard: Lake Baikal.

Lake Baikal is the world's deepest body of fresh water. Inflowing rivers have, over time, deposited nearly 7 kilometers of sediment, but despite this the lake is 1.7 kilometers deep in places. This depth is crucial to the neutrino experiment.

On my first morning in Siberia, we drove across the lake toward the telescope. But before our jeep reached the site, we stopped, gathering with our fellow travelers in the van and another military-green jeep. I knew of the Russians' love for drink, and I was carrying a bottle of fine Scotch whisky to share with my hosts. But I was hardly prepared for the charming ritual I witnessed on the ice—at 10:00 A.M., no less. The white, frozen lake spread for miles around us in every direction except to the northwest, where we were relatively close to shore. Men milled around the vehicles. The subzero temperature seemed to affect everyone differently—some stood bareheaded, others had woolen caps rolled down to the tips of their ears, and then there was Ralf Wischnewski, in his enormous Russian fur cap that looked like a fluffed-up rabbit. A German neutrino physicist who had been working with the Russians at Lake Baikal for twenty years, Wischnewski was the reason I was here. I had met this ruddy-faced man, who was graying at the temples, six months earlier in London, outside the Tate Modern museum, on the south bank of the Thames. We walked over to a Greek pub and discussed the Baikal expedition over chilled lager. It was he who had alerted me to the tradition of bringing spirits to share with the Russians during the winter evenings.

And here we were, except it was still morning. The Russians had planned a traditional welcome drink for Heinze, head of the Moscow office of the Helmholtz Association (the German equivalent of the U.S. National Science Foundation). Kolja Budnev bounded out of our jeep with a bottle of vodka. The jeep's expansive hood became a table, and another bottle of vodka appeared, as did a giant jar of pickled cucumbers. Someone sliced a sausage into circular pieces. As a vegetarian, I was relieved to see that there was some bread. Bright yellow, blue, and red plastic cups were set up on the hood and soon everyone had a vodka-filled cup in their hands. Budnev dipped a finger into his and flicked a few drops of vodka onto the ice—an offering to Burkhan, the great spirit of Lake Baikal. Others did the same before drinking their vodka, and I followed tentatively, not wanting to be drunk as well as jet-lagged (I'd just flown across eight time zones).

We got back into our vehicles and headed toward the neutrino telescope, staying near the northwestern coast. An old railway track, completed in 1904, snakes along the mountainous shoreline. For decades it was an integral segment of the Trans-Siberian Railway. But a hydroelectric power plant built near Irkutsk in the 1950s dammed the Angara River, which flows out of Lake Baikal, and large sections of the track were flooded. New tracks were built beyond the mountains, rendering much of this hundred-year-old engineering feat somewhat obsolete, though local trains still use it. The neutrino experiment outpost depends on these trains to bring in equipment and, more important, groceries. During winter, someone occasionally drives 40 kilometers on the ice to the nearest town for provisions, but the train is the real lifeline, and the only one when the lake is not frozen. About twenty-five years ago, when the experiment started up, the researchers converted two former railway buildings for their own use. The building at kilometer 106 (signifying the distance from Irkutsk) became the shore station — the control room housing all the electronics and some of the staff during their two-month stay every winter. The other building, at kilometer 107, became the canteen and home to the experiment's chief spokesman, Grigory Domogatsky. Trains still stop at the rudimentary platforms of wooden planks, but none that can match the majesty of the Trans-Siberian a century ago.

Four kilometers offshore from Building 106 is the neutrino telescope's ice camp. From the shore, the camp looks like a line of indistinct black dots on the white expanse, towering snow-covered mountains on the far shore acting as a dramatic backdrop. As you get closer, the dots morph into cabins, cranes, jeeps, trucks, an electricity generator, and, closer still, people. On the day I visited, almost everyone was clad in a gray-blue jumpsuit, except for one man who was bare-chested, presumably warm from the work. They were all huddled around a square hole in the ice. A winch had just pulled up one string of the neutrino telescope from the depths of the lake. The top of the string, a collection of hollow balls that act as a large buoy, lay in a tangle on the ice. The rest of the string was still below in the dark waters, and attached to the submerged part of the string were the photomultiplier tubes, to be brought up to the surface for repairs and upgrades.

The Lake Baikal neutrino telescope is made of eleven strings of

photomultiplier tubes—each with a large buoy at the top and a counterweight at the bottom—that float nearly 1.1 kilometers below the
surface (the water here is 1.4 kilometers deep, enough for a building three times as tall as New York's Empire State to sink without a
trace). Smaller buoys attached to the strings float about 10 meters
below the surface. All year round, a total of 228 PMTs watch for the
Cherenkov cones created by neutrinos, monitoring 40 megatons of
water. Each winter, once the ice camp has been set up, the team has
to locate the telescope, the upper part of which drifts slightly over
the course of the year. A diver plunges into the ice-cold water to locate the small buoy fixed to the center of the telescope. Then the
researchers cut holes in the ice above each string (whose positions
they know relative to the center) and attach a winch to the small
buoys to haul up the strings. The team has two months to carry out
any routine maintenance, put the strings back in the water, and get
out before the ice cracks. They have perfected their technique; only
once in two decades of operation did they have a problem retrieving
a string. In 1994, a rusty metal cable broke, severing the buoy from
its string, causing the string to sink to the bottom.

Budnev retrieved it. Diving that deep was out of the question,
but Budnev knew that the string—though its counterweight was on
the lakebed—would still be vertical because of the buoyancy of the
PMTs. What he did next was ingenious. He fashioned a propeller
and tied it to the end of a long rope, dropping the propeller into the
water. The angle of the blades was such that as the propeller sank it
started rotating, making huge circles. Budnev used this simple tool
to sweep the waters below. Soon, the propeller snagged the errant
string, and the team pulled it up. Wischnewski narrated the story of
Budnev's heroics as we stood on the ice, surrounded by the hardiest
physicists I had ever seen. The term "experimental physics" took on
new meaning in this biting cold, which at times dropped to –4°F.

My admiration for these men increased when I saw just how tough
their living conditions were. Wischnewski had warned me in London
that things were difficult. Most of the physicists lived in 10-by-20-
foot cabins, two to a cabin. Others slept in bunk beds in the shore sta-

tion, amid workbenches cluttered with computers, electronics, wires, and cables. They worked long hours, from early in the morning to sometimes well past midnight. There was no running water, which meant no showers for two months. Drinking water was collected each morning from a hole in the ice near the shore—Lake Baikal was pure enough to drink from, the very quality that made it so good for neutrino physics. Toilets were mere pits in the ground, with a wooden cabin around them for privacy. The extreme cold helped control the stench, but it still wafted up when warm urine hit the pit.

There was one luxury here: the *banya,* a traditional Russian sauna. Everyone used it to get clean, but more important, to relieve the stress of working in this bleak, albeit beautiful place. "It is much more important than one can imagine," said Wischnewski of the *banya.* "It is the central part of the weekly rhythm of life here." Naked men sit in an outbuilding, chuck water on hot stones to raise steam, and beat each other with leafy twigs and branches of birch. Then they go out into the freezing cold and pour cold water or rub ice on themselves. It's supposed to bring blood rushing to your skin, open pores, expel toxins, cure disease. Heinze, the German geologist with whom I was sharing a room in a log cabin, swore that he had gone to the *banya* with an upset stomach and had come out feeling fine.

A wicked wind kicked up one evening. Locals call it the *Kultuk,* after the village on the southwestern tip of the lake. It was time for everyone to leave the open ice and head back to the shore station. Once in the shore station, a few of us sat in the kitchen, which looked like a kitchen you would find in a college fraternity house. Years of use had stained the coffee mugs. Dented, discolored pots and pans lay scattered on the shelves. Someone heated up water for tea. I gratefully sat down for a cup, and a can of sweet, syrupy condensed milk materialized. "After a while [here], I start craving chocolate and sugar," said one researcher. "Stress, I guess," he added, as if he needed an excuse for the craving. Another scientist looked at the can wistfully. Condensed milk had been his dream as a child growing up in the Siberian city of Tomsk. "They had this in Moscow," he said, "but not in Tomsk."

Later that evening, I had to walk to the canteen at kilometer 107 for dinner. It wasn't going to be easy. I had turned up on a frozen lake in the depths of a Siberian winter in "European summer shoes," as Wischnewski put it, disbelief in his voice. Now I had to make my way over the lake to the canteen in the dark. I followed Wischnewski. While we were still onshore, I walked along vehicle tracks, where the ice had turned slushy and provided some traction. But on the lake I found walking nearly impossible, my smooth-soled shoes slipping the entire way. After a few days, I learned to find fresh snow for my shoes to grip while walking on the lake, but that night, fear nearly paralyzed me. Fortunately, a jeep pulled up beside us, and Wischnewski, having noticed my plight, asked the driver—Igor Belolaptikov, a tall mustached physicist from the Joint Institute of Nuclear Research in Dubna, near Moscow—to take me to the canteen. I sat with Belolaptikov at dinner and happily accepted a ride back to his small cabin for a chat about neutrinos.

The cabin was a simple affair—one half had two bunk beds, with a long table in the middle. The place bristled with electronics—computers, modems, radios, wires. Belolaptikov shared it with Andrei Panfilov, from the Institute of Nuclear Research in Moscow. We talked for an hour, stopping only for some honey-sweetened tea that Panfilov made. Konstantin Konischev, the bare-chested scientist I'd seen earlier on the lake, now dressed in warm clothes, joined in every now and then, especially when the others struggled with their English.

"My business is the reconstruction of muons and neutrinos," said Belolaptikov, laughing with a childlike joy as he made this disclosure. That reconstruction is tricky business. Hundreds of photomultiplier tubes watch for the flashes of Cherenkov light at the bottom of Lake Baikal. As a neutrino-induced muon races through the water, the light from its Cherenkov cone reaches different PMTs at different times. The skill lies in collecting all the information reaching the PMTs and sifting through it to reconstruct the path of the upward-moving muon. This can then be used to calculate the path of the original neutrino. It is this ability to figure out where a neutrino comes from that differentiates a neutrino telescope from a mere neutrino detector. A telescope has to identify the source of neutrinos in the sky, and the Lake Baikal instrument can do so with an angular resolution of about 2.5°, meaning that it can distinguish neutrinos coming from points in the sky separated by a distance of five full moons. The Baikal telescope—like all other neutrino telescopes that use natural bodies of water or ice—has seen only atmospheric neutrinos. Everyone here is waiting for the day when a high-energy neutrino from outer space will make its presence felt in their little corner of the lake.

Belolaptikov recalled his first neutrino—indeed, the Baikal detector's first. "It was great," he said. "Here, you can see." He leaned over his bunk bed and removed a piece of paper pinned to the wall above. It was a printout of the path of an upward muon, reconstructed from the detection of its Cherenkov cone by the PMTs along the way. "This is the first one—the real, reconstructed one." Teenagers have pinups on their walls; Belolaptikov had this drawing of an event from 1993. And why not? It was the first-ever neutrino seen by humans using a natural body of water as a detector. He and his colleagues had done the reconstruction and put the Lake Baikal detector on the map.

Such scientific scoops must make up for the stress of this job. The expedition, as the Baikal team calls their annual two-month effort, starts with the making of an ice road from Listvyanka to the ice camp. Around mid-February, after the lake's surface has fully frozen, a scouting team sets off from Listvyanka, looking for stretches of ice that seem safe to drive on. Every couple of kilometers, they drill into the ice and remove a core, just to make sure the ice is thick enough. Then they stick a small spruce tree—a recycled Christmas tree—into the hole, and the water freezes around it, leaving the tree stuck firmly in the ice.

Once they make this tree-marked path over the ice, stretching nearly 40 kilometers from the town to their experiment site, the scientists set about locating the submerged telescope. The year I was there, the ice was relatively smooth and the work had been correspondingly easy. But Belolaptikov recounted a particularly harrowing year when the snowfall had been heavy and ice sheets were crashing against each other everywhere, creating massive ridges lined with insurmountably high chunks of ice. One day, Belolaptikov was in the shore station while the others were cutting holes in the ice above the telescope, 4 kilometers from shore. He got a radio call. They were in trouble. "There were big crashes, and very dangerous crashes," Belolaptikov said.

"The ice had shifted in one hour," Konischev, who had been one of the stranded physicists, recalled. Massive cracks, 2 to 3 meters wide, opened up in some places, separating those at the ice camp from the shore station. The team drove along the cracks for nearly 8 kilometers, looking for a way back. "We found one place where we could cross, but we had a bad feeling," said Konischev. "So we called Igor." Belolaptikov scouted the crack from the shore side and found another section that looked safer. "I had to decide something, and I decided that this place was OK," he recalled. "It was really terrible." But Belolaptikov's hunch proved right, the ice held firm, and everyone got safely ashore.

The temperature outside had dropped dramatically. The scientists were almost done with work for the night. Wischnewski had invited everyone to the shore station for a drink—I was to bring my bottle of Chivas Regal. Panfilov joined us, as did Vasily Prosin, a slightly built, extremely fit fifty-nine-year-old physicist from Moscow (he routinely skied the 40 kilometers from Listvyanka to the shore station). Prosin's lanky colleague from Moscow, Leonid Kuzmichev; two Tartars, Vladimir Aynutdinov and Rashid Mirgazov; and a German student named Eike Middell rounded out the group. The whisky was a welcome change from the traditional vodka and Aynutdinov decided that fine whisky should be had on the rocks. He stepped out to get some from the lake, but could not brave the trek down to the shore in the cold. He returned with some sorry-looking lumps snapped off the icicles hanging from the roof of the shore station. During the third round of shots, Prosin and Kuzmichev revealed that they had been fans of Bollywood during their youth, and Prosin and

I broke into a song, *Awaara Hoon* ("I'm a Vagabond"), from a 1960s Hindi film. Wischnewski seemed bemused, clueless as to this cross-cultural exchange.

The whisky finished, Wischnewski brought out a bottle of vodka and poured everyone a shot. As if on cue, each of us dipped a finger into our drinks and flicked a few drops of vodka onto the ground (the gods, it seems, do not care for whisky). The night became fuzzier, despite my gorging on food to keep up with everyone. The conversation flowed, switching easily from Russian to English, and occasionally Italian or German. Wischnewski made the last toast to the Lake Baikal telescope.

The next two days slid by, but even in this short time a rhythm was established. A trip down to the lake in the mornings to get a bucketful of drinking water from a hole in the ice. Then back to the cabin for coffee with condensed milk and honey, making sure to plug the hole in the can of condensed milk with paper to prevent "little animals," as Wischnewski calls insects, from getting in. And then there was my first visit to the outhouse, at 2:00 A.M. Even as I went to sleep the first night, I was worried about this, with temperatures going down to 5°F outside. But nature called, and it meant getting out of my Russian military-issue sleeping bag and into layers of thermals, T-shirts and jackets, thick socks and shoes. A blast of vicious cold hit me as I stepped out of the relatively warm, heated cabin into the naked winter outside. The tire tracks of jeeps that had turned the ground slushy during the day were frozen solid. I was treading on treads. The two-minute walk was its own mini-expedition. Within days, this too became commonplace and the cold just a state of mind. I was walking out in socks and sandals by the end of my stay.

The mornings were ethereal. From my cabin, I could see clear across the lake, its snow-white stillness broken only by patches of bare ice that appeared dark against the snow. I had to remind myself that I was looking at a lake that has more water than the five Great Lakes of North America put together, a lake with a surface area larger than Belgium's. Eighty percent of Russia's fresh water was here. Even at great depths, the lake is well oxygenated, making it one of the most hospitable waters for life.

On a visit to the Lake Baikal museum in Listvyanka, I learned why this body of water is so precious for astroparticle physics. It's because of the voracious crustaceans that live at all depths. Nothing dead or dying lasts more than a few days in this lake. If fishermen leave their catch in the nets too long, the crustaceans invade the fish through their mouths and gills, eating them from the inside out. These critters keep the lake free of dead matter, leaving it unimaginably clear, especially deep down. Murky waters would make watching for muons nearly impossible.

Budnev, who lives in Irkutsk, was on his twenty-sixth neutrino expedition. And he spends time at Lake Baikal at other seasons of the year. "It is a very, very kind water," he said. "It is very important for Russia to preserve Baikal." But that's not necessarily happening. A paper and pulp mill had been operating since the 1960s in the town of Baikalsk, polluting its corner of the lake for decades. And a week before my arrival, protests took place in Irkutsk against an oil pipeline that was to be built along the northern shore. Local politicians and scientists joined hands to fight this plan, which they argued was ill-advised in such a seismically active region. An oil spill would devastate the lake. But no one from Moscow appeared to be listening. This balding physicist, with blue eyes, a stocky build, and thick, rough fingers that belonged to someone unafraid of working with his hands, softened as he spoke of the lake. "If you kill Baikal biology, then there is no way to clean the water," he said. Though he was speaking about the lake's ecology, his manner suggested a greater kinship with these pristine waters.

Besides the lake, Budnev is proud of their detector, which has pioneered the technology of underwater neutrino telescopes. Ironically, the team owes this honor partly to the Soviet Union's invasion of Afghanistan. In the 1970s, an international collaboration that included Americans and Russians was working on building DUMAND, the Deep Underwater Muon and Neutrino Detection Project, in the Pacific Ocean off the coast of Hawaii. Then the Soviet military marched into Afghanistan in 1979, and the partnership started to unravel. The U.S. government threatened to pull funding if the Americans kept working with the Russians, so the teams went their separate ways. The Russians turned to Lake Baikal and the Americans, led by John Learned of the University of Hawaii, stuck to the Pacific. Then, in 1993, disaster struck DUMAND. The researchers had managed to place a string of photomultiplier tubes at the bottom of the Pacific, at a depth of 4.8 kilometers. But soon after it was deployed, the string short-circuited, and the physicists could not communicate with their equipment. Harsh winds and waves kept making things difficult, and finally, in 1995, the U.S. Department of Energy pulled the funding altogether. The Russians went on to build the Lake Baikal telescope, and when it was turned on in 1993, it was the only game in town.

That has now changed, as European physicists have started building similar detectors in the Mediterranean. And a new American-European team moved to the South Pole in the mid-1990s to construct the Antarctic Muon and Neutrino Detector Array (AMANDA), while laying the groundwork for IceCube, the largest-ever neutrino detector. Several German physicists who had worked at Lake Baikal joined the South Pole team. For a few years, Wischnewski, too, split his time between Antarctica and Baikal before committing fully to Baikal. The South Pole detectors are looking for Cherenkov light emitted when muons hit the ice, and IceCube will be watching a cubic kilometer of ice for these ephemeral flashes. The innovations at Baikal—including Belolaptikov's work on reconstructing muons—inspired the early efforts in Antarctica. I would find myself, many moons later, at the South Pole, wondering which of these detectors would make history by seeing the first neutrinos from deep space.

Meanwhile, despite the enormous political chaos that followed the fall of the Soviet Union, the Baikal group has persevered. They still believe that the lake is an extraordinary place to look for neutrinos. While the Antarctic ice is clearer than the waters of Lake Baikal, the water has an edge. Light can travel more than ten times farther in the lake before it is scattered than it can in the ice. Catch the photons before they scatter, and you can tell exactly where they are coming from. Catch them after they have been scattered a few times and it gets harder to work out their original direction. This means that more photomultiplier tubes are needed in the Antarctic ice to gather the information needed to infer the direction of neutrino-induced muons. The Lake Baikal water allows for a less dense array of PMTs. A kilometer-cube detector in these waters would likely be cheaper to build and maintain than the one being built at the South Pole.

Grigory Domogatsky, spokesman for the Baikal project, made this point emphatically one evening, while we were sitting at a table in his cabin next to a roaring fire. Domogatsky smoked filterless Russian cigarettes nonstop. His English was halting, and he had a habit of drumming his fingers on the table or snapping them as he searched for the right words. Despite a rasping smoker's cough that could stop him mid-conversation, he passionately argued that the biggest neutrino detector should be built in Lake Baikal.

The Americans and their European partners were spending $270 million on IceCube, and Domogatsky thought that a tenth of that would be enough to build a comparable detector in Siberia. Besides the advantage of needing far fewer PMTs to detect high-energy neutrinos, Domogatsky pointed out that only a detector in the Northern Hemisphere could see neutrinos from the center of our galaxy.

"But you can see the center of the Milky Way from the South Pole," I said, somewhat puzzled.

"Yes, but not neutrinos," said Domogatsky, with the gentle yet triumphant note of a teacher who has just made a telling point. Of course. Neutrino detectors can see only those neutrinos that come through the Earth. While optical telescopes can study the Milky Way's center from the Southern Hemisphere, only the Baikal detector and the three being built in the Mediterranean—off the coasts of France, Italy, and Greece—will be able to see neutrinos coming from the heart of our galaxy. So physicists need a detector the size of Ice-Cube in the Northern Hemisphere to best observe our galaxy's center—a must for the indirect detection of dark matter.

Domogatsky further argued that Lake Baikal was the best body of water in which to build such a detector, for there are no deep-water currents to contend with, as there are in the Mediterranean. "The currents really don't exist. Lake Baikal is like an aquarium," he said. Besides, scientists in the Mediterranean need ships to lower their strings into the sea and remote-controlled submersibles to wire them up, making the operation expensive. But, Domogatsky sighed, convincing people to work in Siberia during the winter, when the alternative was the sun-soaked Mediterranean, was going to be hard.

Amid this discussion, Domogatsky made tea and laid out a spread of dates, raisins, biscuits, and nuts. He continued smoking, despite his cough. His heavily furrowed face clearly showed the effects of forty years of physics, many of them spent in this hostile place. Now he was looking to pass the baton. The team had just figured out that the telescope they had built so far—eight strings with a total of 192 PMTs within 21 meters of the center and three more strings with a total of 36 PMTs within 100 meters of the center—could form a cell of a much, much larger telescope. Put next to each other, like hexagonal tiles on a floor, such cells could cover a gigaton, a cubic kilometer, of water. All he needed was about $25 million—an order of magnitude less than the money being spent on the Mediterranean projects or at the South Pole.

The fire died. Outdoors, the sun was setting.

"I hope to help start this project," said Domogatsky. "But the work should be performed by younger physicists."

We stepped outside. I took a picture of this grand old man of contemporary Russian physics against the backdrop of his beloved lake and started walking back to the shore station along the railway line from kilometer 107 to 106. The dark silhouette of the canteen's wooden building stood out against an orange-pink twilight sky. A light shone through the dining room doors. Somewhere far behind me, a train sounded a long, low hoot.

There was just one thing left to do. Wischnewski had suggested that my visit would be incomplete without spending a night on the ice. He said I could sleep in one of the cabins at the ice camp. I had agreed. But then he casually mentioned that the ice heaves. Despite the lake's frozen surface, he said, the water beneath is alive and kicking. Sometimes the entire sheet of ice below the camp can jerk and lurch. When that happens the first time at night, you are so scared that you scamper out of your sleeping bag and head straight for open ice. Wischnewski assured me that nothing serious would happen unless a magnitude-6 earthquake hit the region. I wasn't amused.

Walking along the snow-covered railway track, taking care to step gingerly on the wooden railroad ties, I realized that I was afraid. The thought of the ice creaking, groaning, and shifting beneath me as I slept was too much to contemplate. But by the time I got back to the shore station, arrangements had already been made. Someone at the ice camp had agreed to give up his bed for the night. There was no backing out now.

Night came, and Wischnewski and I raced over the ice toward the camp in someone's brand-new Japanese 4 x 4, a far cry from the Russian military jeeps. Flashlights in hand, we walked across the ice for the last few meters; vehicles weren't allowed beyond a certain marker. The sure-footedness I had developed over the last couple of days had disappeared. The cranes and winches looked ominous in the dark. Wischnewski knocked on the steel door of one of the cabins, and a young graduate student named Alexey Kochanov admitted us. He told me not to worry; small earthquakes happened here all the time, he said. Hardly reassuring.

I got into my sleeping bag and again voiced my concern. Kochanov said that he found the sound of ice creaking beneath him relaxing. Obviously he had been here way too long. But then he explained. The creaking means that the ice cover is solid. It is the sound of ice moving in response to the motion of the water beneath. It is only when you don't hear the creaking that you should worry. That's when the cracks are so big that there is plenty of give in the ice. So much give, in fact, that you shouldn't be on the lake.

Suddenly, the ice's protestations were music to my ears. All night it groaned, a sound like nothing I had ever heard, and the sheer variety of groans was tremendous. When they came from far away, they sounded like muffled gunshots or cannon fire or steel doors being slammed shut. The closer sounds were sharper, more like the crack of a whip. Sometimes it lasted a couple of seconds, and sometimes it was right beneath us, a split streaking through the ice below. I finally fell asleep.

At five in the morning, the ice heaved. It was the only significant movement I had felt all night. I couldn't go back to sleep, so I put on my layers of clothing and my socks and shoes and went outside. It was still dark. The ice did not open to swallow me. Thin cracks criss-crossed the surface. You could tell that the new fissures had formed in the night, because they had not yet been covered by snow. Every-where, thinner cracks had refrozen, forming intricate patterns. In the feeble beam of my flashlight, these flowerlike designs appeared in relief against the darkness below. A generator hummed nearby, but it did not ruin the quiet and stillness of this immensely cold place. Moonlight played with the ice, barely hinting at its thickness and the vast expanse of bone-chilling water beneath. Scorpio's tail was visible next to the moon, and overhead was Ursa Minor. On the far shore, embers of a forest fire glowed on the slopes of the Khamar-Daban range.

I got back into the cabin and into a bed warmed by electric heat-ers. Somewhere deep below, a cone of bluish light raced upward through the cold water. A neutrino had traveled from some distant part of the universe, escaped collision with every bit of matter on its way, and gone right through the center of the Earth, only to collide with a molecule of water in Lake Baikal and disappear in a flash of light. Will the artificial eyes below see such a flash one day and know that they have seen, for the first time, a ghostly particle from the cen-ter of the Milky Way?

If so, it could help unravel the mystery of dark matter and illumi-nate the makeup of about a quarter of the universe. To understand how cosmologists are tackling the rest of the unknown, I traveled to a place that is the antithesis of Siberia: a parched desert in the Chil-ean Andes.

4 · THE PARANAL LIGHT QUARTET

T HE FIRST THING that struck me as I flew into Antofagasta was that the city had few trees. Everything in the landscape was a shade of brown or gray, from dark, rusty rocks to pale, ashen silt. The dust-laden trees lent some semblance of green, but even they seemed forlorn and entirely out of place in this desolate land. Antofagasta is a port city on Chile's northern coast, on the edge of the Atacama Desert. To the west lies the Pacific Ocean, which today looked sullen beneath a cloudy sky. To the east are the barren hills of the Cordillera de la Costa, which rise abruptly and dramatically. Antofagasta is, like most of Chile, a sliver of land wedged between mighty mountains and an immense ocean.

That's how a middle-aged, stubbled, and very tired-looking Chilean had put it a few hours earlier, in the Santiago airport, animating his halting English by holding his forefinger and thumb an inch apart to indicate just how thin his country is. We were both waiting for the flight to Antofagasta. I told him I was headed to Cerro (Mount) Paranal to see the observatory. He had heard not just of Paranal but of all the other major Andean observatories in Chile—Cerro Tololo, Las Campanas, Chajnantor, and La Silla (I pronounced the *l*'s and he gently corrected me). I was surprised to hear that a mining executive—he told me he operated open-pit copper mines near Antofagasta—knew about these telescopes. "My company spends a lot of money on lights," he said, by way of explanation, miming an upside-down cone with his hands to show how the facility's lighting was designed to shine downward instead of toward the sky.

The country's dark skies are a national—indeed, an international—treasure. Chilean laws limit light pollution, even if it means that everyone spends a little extra to abide by the law. "People don't mind," said my travel companion. "It's important for Chile to have the observatories."

I couldn't wait to see those night skies. Friends had talked about the sheer number and intensity of stars, partly the bounty of the southern skies and partly because of the paucity of people and lights in many parts of Chile. And then there is the Atacama Desert, one of the driest regions of the world. It lies on the leeward side of the Andes, which act as a formidable barrier to the moisture-laden winds from the east. Some parts of the desert haven't seen rain for years; some haven't seen any in recent memory. Couple the dryness with the desert's high altitude and the stillness of the air above it and you have some of the best "seeing," as astronomers say, in the world. There's little of the atmospheric turbulence that haunts even the best of telescopes.

Cerro Paranal is a 2,635-meter-high mountain in the Atacama Desert, about 120 kilometers south of Antofagasta and barely 12 kilometers inland from the Pacific coast. Despite a clear line of sight, the ocean is rarely seen from the mountain, as it is often obscured by a thick layer of clouds, the kind you would normally see from an airplane at 40,000 feet. These low-lying clouds are exactly what make Cerro Paranal a perfect spot for astronomy. The cold Humboldt Current, flowing northward along the Chilean coast, creates a strong inversion layer, pulling the clouds down to well below the summit of Paranal, not just reducing the moisture in the air around its telescopes but also creating nearly 350 days of clear skies. The dry air above Paranal — low in water vapor — allows light from outer space to reach the telescopes without being absorbed by the atmosphere. Nowhere else on Earth do these climatic features — a high-altitude desert and an ocean-induced inversion layer — come together as they do in the Atacama.

But what delights the astronomers about the sky is also what makes the region so hostile to life. Water is an extremely scarce resource, and only recently have the native Indians who live along the coast started harvesting it, using an ingeniously simple method they developed with help from scientists. They trap the fog that rolls in from the ocean with large polypropylene nets, causing it to condense, drip into gutters, and flow into cisterns. It's the same trick performed by the coastal redwoods in California, whose numerous branches and dense array of needles are such effective traps that on particularly foggy nights the water falling on the ground beneath a giant redwood can rival a rainstorm.

In 1991, the European Organisation for Astronomical Research in the Southern Hemisphere (ESO) chose Cerro Paranal to build its artfully named Very Large Telescope, or VLT. The work started the same year and continued for a decade. Today, the VLT is one of a handful of giant telescopes around the world that are peering more than halfway across the known universe in an effort to unravel one of cosmology's newest mysteries. Our universe has begun expanding at an increasing rate, and cosmologists attribute this acceleration to something called dark energy. On my trip to Paranal I wanted to start making sense of this discovery, which has shaken up cosmology — and, indeed, all of physics. Dark energy, whatever it is, is thought to make up about 73 percent of the universe. That's an awful lot we don't understand.

It took more than twenty-four hours of flying and waiting in airports to get from London to Antofagasta. On the longest leg of the journey, a more than 6,000-mile flight from Madrid to Santiago, the plane crossed the equator into the Southern Hemisphere. We entered South American airspace somewhere above Fortaleza, Brazil, and then cut diagonally through the heart of the great continent, heading southwest toward Santiago. I tried hard to stay awake during the night flight over the Amazon, to see something I had only read about: the lightning of violent thunderstorms that rage over the steaming jungles. But the jet encountered little turbulence and I slept right through, only to wake up just as we were about to land. It was early morning. I could see the foothills around Santiago, their innocuous brown-green mounds accentuating the snow-streaked jaggedness of the Andes in the distance.

Many hours later, when it was twilight in Antofagasta, my excitement at being in Chile was replaced by a strange dread. I walked from my hotel-by-the-sea into the city center and saw little but wide, empty roads and shuttered shops. I didn't know it then, but this was a typical Sunday evening in this Catholic country. There were very few people in the central square—mainly armed Chilean police. The buildings were disparate—old, rundown structures amid spanking new stores painted in bright reds and blues and fluorescent greens and yellows, all set against the dusty dryness of the land. The only shops that were open were well-lit pharmacies. I later learned that Chileans are avid self-medicators, hence the profusion of stores selling medicines. That night the empty streets felt alarming.

The next morning, I left my hotel and walked across the railway line that connects Antofagasta to Bolivia, the tracks rising from sea level to an altitude of some 15,000 feet. On the other side of the tracks was ESO's Antofagasta office, where I waited for a bus to Paranal. The visitors' room was adorned with pictures of galaxies and nebulae taken by the various ESO telescopes in Chile. The most striking of these was the Sombrero galaxy, the galaxy Vesto Slipher had studied in 1914 and found to be moving away from us at a speed of 1,100 kilometers per second. Slipher, who used a 24-inch refractor for his observations, could hardly have imagined the clarity of this image. The core of the Sombrero bulged with old, red stars. Its halo looked bluish, with faint blobs that weren't individual stars but globular clusters of a million stars each. A dark dust lane circled the core, sharply silhouetted against the luminous heart of the galaxy and looking like the rim of a Mexican hat.

It was Slipher's observation of the Sombrero and other galaxies that laid the foundation for Edwin Hubble's discovery of an expanding universe. This extraordinary finding in turn led to the theory of the big bang. The theory accounted for the expansion, but it also suggested that as the universe ages, gravity should eventually gain the upper hand and cause the rate of expansion to diminish. But in less than a century after Hubble's work, huge telescopes around the world, including ones in Chile, brought about another upheaval in our understanding of the universe. In 1998, cosmologists announced that the expansion of the universe, rather than slowing down, is speeding up. Nothing in our theories comes close to explaining what might be responsible. The likeliest explanation is that the fabric of space has an inherent energy — called dark energy — that is creating a repulsive force strong enough to counter gravity.

However, the journey from an expanding universe to one filled with dark energy could not have happened without several key intervening discoveries. The first was the accidental sighting of the radiation left over from the early universe. Called the cosmic microwave background (CMB), it was stunning proof, finally, that the universe had begun in a fireball.

James Peebles started wondering about the existence of the cosmic microwave background in 1965. Ensconced in his high-ceilinged, book-lined office on the leafy campus of Princeton University, the tall, slightly stooped Peebles comes across as a kind, elderly man full of effortless humility and self-deprecating humor. Peebles singlehandedly nurtured (some say gave birth to) the field of physical cosmology, which asks fundamental questions about the formation and evolution of the universe from observations of its large-scale structure. "Of course, I wasn't alone," said Peebles. "But I played a large part. I have an easy excuse. It was a very small field in the 1960s, and a field that was unplowed."

Not only was the field of cosmology unplowed, but many physicists didn't even want to rake it, afraid of its speculative nature. Was the universe in a steady state? Did it begin in a big bang? Though the idea of an expanding universe had been confirmed by astronomers, the notion of a big bang was still hard to stomach, for the laws of physics broke down when addressing the supposed beginning—a state of seemingly infinite density. So from the late 1940s to the early 1960s, scientists continued to wrangle over the nature of the universe. English cosmologists Fred Hoyle, Hermann Bondi, and Thomas Gold promoted the theory of a steady-state universe—one that has no beginning and will have no end. To account for Hubble's expanding universe, the theory relied on the continuous creation of new matter, which would keep the matter density constant even as spacetime grew. A philosophically pleasing scenario, the steady-state universe had its share of followers.

However, its opponents, most notably George Gamow, were fierce in their criticism. "It is not surprising that the Steady State theory is so popular in England," said Gamow, "not only because it was proposed by three (native-born and imported) sons H. Bondi, T. Gold, and F. Hoyle but also because it has ever been the policy of Great Britain to maintain the *status quo* in Europe." The war of words spilled over to popular writings and even the airwaves. Then, in 1949, during a BBC radio program titled *The Nature of Things,* Hoyle took on the theory that argued that the universe had a beginning, using the term "big bang" in a moment of derision. It was the first time the name was used, and it stuck.

The matter could not be resolved without additional experimental data. "[Cosmology] had a bad reputation for being highly speculative, more a debating club than a science," said Peebles. "When I started working in cosmology, the faculty was always supportive. But I got the impression that their feeling was 'Better you than me.'"

Peebles came to Princeton to study theoretical particle physics, but Robert Dicke, an experimental cosmologist, dragged him aside and suggested that Peebles ought to work with him rather than with "those particle theorists." The young Peebles consented and became the theorist in a group of experimentalists. Dicke, who had invented an advanced radiometer during World War II, wanted to use the technology to measure any radiation that might be left over from the big bang. "Why don't you think about the theoretical implications?" Dicke asked Peebles.

"Those deeply profound words set my career," Peebles told me, occasionally closing his eyes during these reminiscences, as if he were being transported back in time.

Peebles ended up rediscovering something Gamow had already figured out in the 1940s. Gamow's work didn't make a splash when it was published and was soon forgotten. But he had realized, as did Peebles, that the universe today must be awash in fossil radiation from the big bang. The infant universe would have been a combination of photons, electrons, and hydrogen and helium nuclei. All matter existed in a hot, ionized state for the first few hundred thousand years, which prevented the photons from going anywhere, as they kept getting scattered by the free electrons. Then as the universe continued to expand, it cooled enough for the electrons and nuclei to combine into atoms, and suddenly, in what is known as the epoch of recombination, the universe became transparent to photons. Redshifted all the way into microwave frequencies because of the expanding universe, these photons should now have a temperature of just a few degrees above absolute zero. Dicke and his team set out to look for this cosmic microwave background, unaware that in a nearby town two radio engineers had already seen it.

The story of how Arno Penzias and Robert Wilson beat Dicke's team to the discovery is now legend. In the early 1960s, the two physicists, working at Bell Labs in Holmdel, New Jersey, had taken charge of a special horn antenna that until then had been used for radio communications with satellites. The duo started using the antenna to study radio emissions from the Milky Way and found that the antenna was picking up a faint signal that seemed to have all the characteristics of electronic noise. The direction of the antenna didn't matter; the signal was always there. Stories abound about their attempts to clean the antenna of all sources of noise, including pigeon droppings. Nothing they did got rid of it. The signal was real.

Word about the Princeton team's work reached Penzias and Wilson via a colloquium in which Peebles described their effort and the theory behind it. Penzias and Wilson realized the significance of their discovery, and Penzias called Dicke to inform him of the strange signal. Peebles remembers that fateful phone call: "We were having

lunch in Bob Dicke's office. He answered the phone and started talking to Arno Penzias." Dicke put the phone down, turned around to
his team members, and said, "Boys, we have been scooped."

Peebles's reaction was one of "great relief." Finally, the debate
could be settled. The steady-state universe had to give way to the big
bang, for here was proof, however tentative, of the predicted fossil
radiation. The discovery proved so momentous that Penzias and Wilson received the Nobel Prize in 1978. Dicke was overlooked, which
still rankles Peebles. "I remember being shocked that Dicke didn't
get it," he said. "I remember seeing him that day. Very characteristic
of him—he refused to say anything, just shrugged his shoulders."

Was Peebles resentful of not getting the Nobel himself? No was
the prompt answer. "This was not my prize, this was Dicke's," he
said. "Perhaps I felt a little chagrin, because I was the medium for
alerting Penzias and Wilson to the Princeton work."

The discovery of the CMB cemented the notion of a big bang.
But for all its elegance, that theory had thrown up some intractable
problems. Soon after the CMB was discovered, Dicke went to Cornell to talk about an aspect of the big bang that was seriously both-
ering theorists: the flatness problem. According to measurements at
the time, the density of matter seemed almost equal to the so-called
critical density—the density that was needed to make the curvature
of the universe flat. (A flat universe is one in which two parallel lines
remain parallel all the way to infinity, whereas in a "closed" universe
the lines will converge. In an "open" universe, two parallel lines will
eventually diverge. The two-dimensional surface of Earth is closed:
Two lines that start off parallel at the equator will touch at the poles.
A saddle provides an example of an open surface on which two parallel lines will diverge).

As Dicke pointed out in his talk, if the matter density had been
greater, then the universe's curvature would be closed, and if it had
been lower, the universe would be open. But the universe seemed to
be flat, meaning that the ratio of the actual matter density to the critical density, which is denoted by a parameter called Omega, was very
close to 1. And for today's universe to have Omega anywhere near

1, its value just one second after the big bang would have had to be exactly 1 to a precision of about fourteen decimal places. Nothing in the laws of physics suggested why Omega should be so finely tuned in the early universe, when it could have had any arbitrary value.

In Dicke's audience was a young postdoc named Alan Guth. He was a particle physicist who had no interest in cosmology. But something about the talk tickled his fancy and set him on a journey that would solve the big-bang theory's most frustrating problem.

I went to see Alan Guth at the Center for Theoretical Physics at MIT one fall day at the appointed hour of 10:15 A.M. His office was closed. Guth's secretary came by to drop off some mail. "Oh, I have to tell you something," she said. "He's always fifteen minutes late." As if on cue, Guth came walking down the corridor at exactly 10:30, a stocky, slightly hunched man, his trousers wrapped in fluorescent-yellow ankle bands, bicycle helmet in hand. He apologized and we went into his office, which was surprisingly neat. You would never know that not long ago it had earned the dubious honor of being the Boston workspace most in need of a makeover. I asked Guth about the transformation. He laughed and said that as a prize he had won the services of a company that helped him clean up and get organized.

The conversation quickly turned to 1979, when he and Henry Tye were postdocs at Cornell University. Tye was interested in strange hypothetical particles called magnetic monopoles, which are essentially magnets with only one pole, unlike regular magnets, which always have a north and a south pole. They are analogous to charged particles with either a negative or positive charge. Magnetic monopoles are allowed by James Clerk Maxwell's theory of electromagnetism. "The theory looks like it is made for them," said Guth. "Yet none are known to exist, and that has always been something of a puzzle." Tye wanted to work out how many magnetic monopoles, if any, were produced in the big bang. Guth thought it was crazy for a particle physicist steeped in mathematics to take on cosmology, a field that was still far from precise.

Dicke's talk wasn't the only one to change Guth's mind. Another

was a lecture by Steven Weinberg, in which he talked about baryogenesis: the production of baryons—that is, the bulk of the normal
matter—after the big bang. Weinberg was using so-called grand unified theories (GUTs), which deal with the universe when it was just
10^{-35} seconds old, a time when three of the four fundamental forces
of nature (all except gravity) were acting as one and indistinguishable from each other. Guth decided that if Weinberg was taking the
big bang seriously, it must be important. This was before Weinberg
had won a Nobel Prize for his work in particle physics, though he
"still looked like a Nobel Prize winner—even then you could tell,"
said Guth.

So Guth and Tye started seriously wondering whether the big
bang could have produced magnetic monopoles. They focused on
what might have happened as the universe underwent a transition in
which the three unified forces separated into two. This is called the
GUT phase transition and was caused by the cooling of the universe
as it expanded. Their calculations showed that magnetic monopoles
should have been produced in copious amounts during this period.
This was a problem, as not a single monopole had ever been detected. But soon the duo found a way of suppressing the production
of monopoles during the phase transition, explaining why we don't
see them today.

It had to do with a phenomenon called supercooling, a process
that can be observed if you carefully freeze water. Under certain conditions, the temperature can drop below 0°C even as the water remains liquid. The slightest disturbance, however, can initiate the
phase change from supercooled water to ice. Guth and Tye reasoned
that the universe could have similarly supercooled as it expanded.
If so, it would have remained in a grand unified state even though
its temperature was low enough for the forces to have separated.
And their calculations showed that when the phase transition eventually occurred, magnetic monopoles would not have been produced
in such large numbers. So far, so good. They had seemingly solved
the magnetic-monopole problem. There was, however, one small issue. "While we were working together," said Guth, "we never wor-

ried about the question of how this supercooling might affect the expansion of the universe."

That was 1979. Guth left Cornell for a one-year stint at the Stanford Linear Accelerator Center (SLAC). Tye embarked on a trip to China. Late one December night, alone at SLAC, Guth finally tackled the question of how supercooling might influence the expanding universe. The calculations, once identified, were easy to do. "The conclusion is that it has a dramatic effect on the expansion rate of the universe," said Guth. "The universe grows exponentially."

This exponential expansion had to do with the properties of the vacuum of spacetime during the phase transition. Every bit of vacuum has energy, and the energy density of the vacuum is greater before the GUT phase transition than after. In the supercooled state, the vacuum is stuck in a state of higher energy density known as false vacuum. Einstein's general theory of relativity shows that such false vacuum will produce a repulsive force, causing spacetime to expand. This creates more false vacuum, hence more repulsion. So with every tick of the clock, the supercooled universe doubles in size, expanding exponentially. Eventually, the false vacuum starts undergoing a phase transition. Bubbles of the new vacuum form—analogous to crystals of ice in supercooled water—and collide with one another. Soon the entire universe enters the new phase.

Guth stared at his calculations, aware that he had stumbled on something special. "I realized the same night that that would answer Dicke's flatness problem," he told me. The exponential expansion—which Guth named inflation—would drive the universe toward an Omega of 1. The Omega of the universe prior to inflation did not matter. No fine-tuning was required, for inflation had expanded the universe to such an extent that it erased any and all of that initial history. An excited Guth raced to his office the next morning ("I broke my speed record biking from my apartment to SLAC") to complete the calculations. When he convinced himself that he had solved the flatness problem, he wrote the words "spectacular realization" in his notebook and drew a double box around it.

Inflation soon became the rage. A major reason for its quick acceptance was that it did not change anything about big-bang theory. Inflation would have occurred before the start of any of the processes of interest to cosmologists, such as the nuclear synthesis of elements. After inflation ended, the universe would continue expanding, but at the much steadier rate determined by Hubble. Guth became an instant hit on the lecture circuit, embarking on a five-week tour of universities in 1980. Job offers poured in. Six, seven, eight—he barely remembers now. What he remembers is that MIT, his alma mater, had no opening; he hadn't applied there for a job, though it would have been his first choice. Toward the end of his tour, Guth found himself in Maryland, eating dinner at a Chinese restaurant. He broke open his fortune cookie, and the message said something like "Exciting opportunity awaits if you are not too timid." Goaded into action, Guth called MIT and asked for a job. "To my delight," he said, "they called me back in twenty-four hours and offered me a position"—as an associate professor of physics. He accepted, and just in time, too, for as he himself would soon realize, his version of inflation was not quite right.

Still, the big-bang theory was now on firmer footing. The cosmic microwave background proved that there had been a fireball. Inflation plugged the holes in the theory. Cosmologists were now looking for more confirmation. If the theory was right, surely the expansion of the universe, nearly 14 billion years later, was slowing down. They were in for a shock.

The trip from Antofagasta to Cerro Paranal takes a little over two hours. A plush Mercedes bus took us south along Antofagasta's oceanfront before turning east and heading for the hills. The land morphed with stunning rapidity, cityscape giving way to what looked like giant sand dunes; they were not dunes, however, but rocky hills that grew with each passing mile. A huge cement factory covered in a haze of its own making added to the sense of passing through extremely dusty land.

The road from Antofagasta merged into a section of the Pan American Highway, which is the name for the tens of thousands of miles of roads that connect Alaska to the tip of South America, barring a few gaps in the hostile stretch of jungle between Panama and Colombia. After a short ride on the highway, our bus entered the Old Pan American Road, a stretch of compacted gravel that made for a harsh ride even in our luxurious Mercedes. It was hard to tell where the edge of the road ended and the desert began.

Eight years earlier, the same road had shuddered under the weight of a convoy that carried the first of the four gigantic 23-ton mirrors that make up the Very Large Telescope. Each mirror, made of a high-tech glass-ceramic called Zerodur, had been cast in Germany from 45 tons of the molten material. The glass-ceramic had been poured into a mold and spun. As it cooled, centrifugal forces combined with the mold's curvature to bring the material close to the desired shape. German engineers then ground the blank before sending it to France, where it was polished relentlessly for two years, so that the 50-square-meter surface came to within 0.00005 millimeter of the required curvature. Imagine sanding down a city like Paris so that the only bumps left were about a millimeter high—such was the precision of the polishing. In 1997, this mirror, 8.2 meters in diameter, became the largest of its kind ever manufactured, so large that the steel structure built to support it weighs 10 tons. Engineers had pushed the very limits of technology.

To this day, the VLT mirrors remain among the largest mirrors ever made from a single piece of glass. The bigger a primary mirror gets, the thicker it has to be, to withstand the deformations caused by gravity as the telescope moves from zenith to horizon. But the thicker the mirror gets, the heavier it becomes; for mirrors bigger than about 8 meters the weight becomes unmanageable. While bigger telescopes have been built since the VLT, none have surpassed its mirror size significantly without resorting to combining smaller pieces of glass, and nowhere else can you find a quartet of 8.2-meter mirrors in such close proximity.

The first of the four gigantic mirrors came from France to Antofagasta in December 1997, after a month-long voyage by ship. Instead of the two hours or so it would take me, it took a massive trailer more than two days to reach the mountaintop. You could have walked up alongside it. The fragile load was treated with extreme care; at times, road graders smoothed the path in front of the trailer. In fact, the operation was so delicate that the ESO staff first did a test run of the entire process using a concrete dummy mirror weighing all of 23 tons, and only when they had worked out all the kinks did they transport the real mirror. Astronomers must have held their collective breath until the fourth and final mirror made it to the mountaintop in 2000.

From the bus window, I caught my first glimpse of the telescopes—four white blobs on a peak. The white was the intense light reflecting off the metal enclosures of the telescopes. Until then, the landscape had been an unbroken litany of brown hills. Any sign of life or human activity—power lines, billboards, trees—had long since vanished. Amid the barrenness of the Atacama Desert glistened the giants of the Paranal Observatory.

The final stretch of road was as smooth as an airport's tarmac and climbed steeply upward. We had entered a special 725-square-kilometer region that Chile had donated to ESO. The bus let us off at the base camp. We were at 2,360 meters, and the cold air stung. I walked toward what looked like an enormous dome buried in the ground. It was a subterranean facility that had been built in a natural depression on the flanks of Cerro Paranal, a residence for astronomers, engineers, and others working there. It would be my home for the next few days.

A long ramp sloped down toward the dome, at the end of which was a set of surprisingly heavy steel doors that led into a small dark passage. Another set of heavy doors stood at the opposite end; it was like a submarine's air lock. When I walked through the second set of doors, the smell of wet earth enveloped me. After two hours of the dry Atacama air, the humidity inside the dome was overpowering—a welcome sign of life in an otherwise hostile land.

Inside was a profusion of green; a small tropical garden—cacti, ferns, bougainvillea, palms, banana trees—spilled downward into a cavernous facility. Sunlight poured in through a translucent ceiling.

A giant pole extended from the floor, capped by a huge furled umbrella that could be unfurled at night to keep the light from the dome
from escaping into the night sky. A ramp curved in a gentle arc along
the edge of the garden toward the reception area below. Through
the tangle of green, I saw a glimmer of blue—a swimming pool, the
source of the astonishing humidity within the dome.

In a desert devoid of the usual rhythms of life, the dome offers
residents a sanctuary with a rhythm of its own. At night, the building shuts itself to the night sky, its metal and plastic dome completely shrouded by the giant umbrella. Every blind on every window
in every room is pulled shut, and not a sliver of light escapes. All this
is for the benefit of the four great telescopes, which are looking for
photons from galaxies billions of light-years away. Come daybreak,
the dome opens again, as a flower to sunlight.

My guide was ESO's press officer, Valentina Rodriguez, a pe-
tite Chilean with dark brown hair and sparkling brown eyes. After a
sumptuous dinner in the dome's canteen, she drove us up the nearly
2 kilometers of road leading from the base camp to the summit of
Cerro Paranal.

The summit, which is about 275 meters higher than the base
camp, supports a flat platform, built by razing nearly 300,000 cubic
meters of rock and soil from the mountaintop. On the platform stand
the four immense metal enclosures. They are rectilinear, with nearly
flat roofs, a departure from the usual rounded domes of telescopes.
They rise like sentinels guarding the secrets of the cosmos. Standing
beside them made me realize how towering they were (nearly twenty
times taller than the average human being), something that wasn't
obvious from the base camp.

The sun was going to set soon. Engineers were getting ready to
open the enclosures and aim the telescopes at the sky. They would
ensure that everything was in working order before handing the telescopes over to the astronomers. Valentina had brought me here to
see the opening.

We entered through a heavy steel door, and the instant the door
shut behind us it was pitch-black inside. The enclosures are hermeti-

cally sealed during the day, both to keep the dust out and to keep the optics cooled to the forecast temperature at sunset, so that when the enclosures open, the mirrors will not be affected by a sudden drop in temperature. Just then, an engineer walked in and flipped a switch, and a few lights flickered on. Valentina led me up a metal staircase onto a circular platform, and I got my first glimpse of one of the famed 8.2-meter mirrors.

The primary mirror is 18 centimeters thick. Because of its weight, the mirror's precise shape can warp when it is tilted, so 150 actuators, upon which the mirror rests, continually push and pull at least once a minute to ensure that the optimal curvature is maintained. More impressive than the actuators are the clamps around the edges of the mirror, which can, at a moment's notice, lift the entire mirror, all 23 tons of it, off the actuators and secure it to the telescope's support structure in case of an earthquake (moderate quakes, of less than 7.75 Richter, are not uncommon here, thanks to the ongoing collision of the Nazca and South American plates). The entire telescope is designed to swing during an earthquake, and securing the primary mirror prevents it from rattling against the metal tubes that surround it.

The platform we were standing on runs all the way around an inner structure that holds the telescope. Everything in this building except the platform can move. The telescope moves to track celestial objects, and the enclosure moves so that the opening in the roof is right above the telescope. The primary mirror is supported on a massive bowl-shaped truss, which itself is at the bottom of a tubular truss that points straight up. At the top of the tube is the secondary mirror, made of extremely lightweight beryllium. Unlike the 100-inch telescope on Mount Wilson, which originally had a flat secondary mirror that reflected the light out of the tube at the very top, the secondary mirrors in modern telescopes are curved. When beams of light from distant stars and galaxies strike the primary mirror, it focuses the light at the secondary mirror, which then reflects it back to the very center of the primary. At the center of the primary is a hole, and positioned below that hole is another mirror, which can deflect the beam toward a camera or a spectrograph.

The entire truss assembly that supports these mirrors can pivot on a gigantic fork, so that the telescope can either point straight up or swing down nearly to the horizon. The fork itself rotates on its vertical axis to point the telescope in any direction along the horizon.

The whole structure towered over us, eerily lit by the few lights inside the enclosure. Suddenly, I felt the platform start to move, though it was actually the telescope rotating on its vertical axis. It was so quiet that I hadn't noticed—a remarkable achievement for an instrument that weighs over 400 tons. I had been in older telescope domes before, and the telescopes usually groaned and creaked like mighty beasts. This one was silken smooth, because the entire assembly was supported by oil-filled hydrostatic pads. The only noise so far had been the hiss of the hermetic seals opening, letting in the warm glow of the setting sun. I felt the full impact of the mirror's size when the top of the tube swung down toward us, so that the mirror was now facing us and reflecting the very structure it was a part of. It made me dizzy. I stepped back reflexively.

"Watch your back," said Valentina.

Sure enough, the enclosure had started moving, too. It stopped after a few minutes, and there was now considerable clanging, as a slot in the enclosure opened, and slats and louvers designed to control wind flow into the enclosure slid and swiveled. "For me, it is like a dragon waking up," said Valentina. We watched until the telescope had swung back up to point at the sky through the opening, and then the engineer herded us out. The telescope belonged to the astronomers for the night, and we headed down to see them.

The control room was located below the main platform. There I spied Chris Lidman, a member of one of the teams that had discovered dark energy in the late 1990s. He was standing in a large cubicle filled with computer screens. There were four such cubicles, and each contained computers that controlled one of the four telescopes, which the astronomers refer to as UT1, UT2, UT3, and UT4. They have more evocative names—Antu, Kueyen, Melipal, and Yepun (for Sun, Moon, Southern Cross, and Evening Star, in the language of the Mapuche Indians)—though I didn't hear those names used inside the control room.

Lidman, a compact, wiry man with wispy brown hair, clad in a long-sleeved black turtleneck and brown slacks, had just arrived in Paranal for ten days of observing. An Australian, he had been working with telescopes in Chile for eleven years. I introduced myself to him and his colleague Markus Hartung, an Austrian astronomer also resident at Paranal.

"You should show him this," Hartung said to Lidman.

I turned to look at the computer. On the screen flickered five bright blobs, one at the center and the four others positioned at the corners of an imaginary cross. The image was coming from UT1.

"This is a very famous object called the Einstein Cross," said Lidman. It was caused, he said, by a phenomenon predicted by Einstein's general theory of relativity—the bending of light from a distant object by the gravitational field of a massive galaxy in the foreground. The blobs were actually a quasar, which is a giant galaxy with a supermassive black hole at its center. Prodigious amounts of light had left the quasar nearly 8 billion years ago, and the light's wave front was now spread over a sphere with a radius of 8 billion light-years (1 light-year = 9.5×10^{12} kilometers). Now, imagine Earth, an insignificant blue dot on the surface of that sphere—and consider that few photons from the quasar must have reached our planet and fewer still must have settled on UT1's 8.2-meter mirror, itself a tiny silver dot on Earth's surface. If it hadn't been for that massive galaxy between the quasar and the VLT, the quasar might have been too faint to see. The galaxy was shepherding the quasar's light toward us, and those photons, or their electronic counterparts, were now shimmering on the screen.

Testing Einstein's general theory of relativity has not always been this straightforward. In fact, the first confirmation of general relativity required expeditions to Africa and South America, in search of the perfect site from which to observe a total solar eclipse. Einstein had predicted the amount by which the gravity of the sun would bend the light from a star toward us. Normally, the bending of starlight by the sun is impossible to detect, because of the sun's blinding brightness, but a total eclipse reveals the stars close to the sun, and by comparing their positions during an eclipse to their positions in the night sky, one can find out just how much the starlight is being bent by the sun.

When Einstein published his general theory, in 1916, he over-turned long-cherished Newtonian notions of gravity. No longer was it some kind of force exerted by one celestial object on another. Grav-ity, according to Einstein, arose as a result of the curvature of space-time. Four-dimensional spacetime was not just a static backdrop to the motion of stars and planets, it was an active participant in the celestial drama. Any object, however small, caused spacetime to curve in a manner proportional to the object's mass, and everything in the vicinity of the object was now constrained to move along that curved spacetime. We can best visualize this by imagining a heavy ball placed on a taut rubber sheet. The ball dents the rubber sheet, and if a smaller ball starts rolling along the sheet, it has to follow the curvature forced upon the sheet by the heavier ball. If the smaller ball is too far away and the rubber sheet around it isn't affected by the heavier ball, the smaller ball will move as if the heavier one did not exist. The rubber sheet is analogous to spacetime, and the balls are analogous to planets, stars, and galaxies. Gravity, then, is the cur-vature of spacetime.

Light, like stars and planets, also has to follow the curvature of spacetime—for where else can it go? Einstein predicted that light from a star passing close to the sun would bend ever so slightly to-ward the sun. In a now legendary experiment, British astronomer Arthur Eddington traveled to the island of Príncipe, off the coast of West Africa, to observe the position of stars during a total so-lar eclipse on May 29, 1919. That year, the eclipse could be viewed all along a stretch of the Atlantic, from Brazil to West Africa. Help-fully, the eclipsed sun was passing in front of the Hyades, a bright cluster of stars in the constellation Taurus. Eddington photographed these stars during the eclipse and claimed that the sun was bending their light. Another team of astronomers, who had traveled to So-bral, in northern Brazil, to study the same phenomenon during the same eclipse, corroborated the findings. On September 27th, Einstein wrote in a postcard to his mother, ". . . joyous news today . . . English expeditions have actually measured the deflection of starlight."

Then, on November 6, 1919, at a special meeting of the Royal Astronomical Society and the Royal Society of London, the chair of the meeting declared, "This is the most important result obtained in connection with the theory of gravitation since Newton's day." Over time, it would become clear that technical problems encountered by the two teams meant that errors in measurement were too great for the result to constitute absolute proof, but the world had already anointed a new king of physics; newspapers plastered Einstein's face on their front pages and turned him into an instant celebrity.

Einstein's theory has resulted in another fascinating discovery. As noted, when the sun's gravity warps spacetime, it creates the equivalent of a gigantic convex lens, and as light from a distant star passes by the sun, it is bent by the lens toward the observer in front of the lens. In 1924, the Russian physicist Orest Chwolson used Einstein's general relativity to propose the existence of such "gravitational lenses"—lenses created by the gravity of stars. Twelve years later, Einstein himself quantified the properties of such lenses but concluded that "there is no great chance of observing this phenomenon" for stars other than the sun. What even he did not realize was that there were powerful gravitational lenses being formed by objects more massive than stars. In 1937, astronomer Fritz Zwicky made a visionary claim. Extragalactic nebulae (or galaxies, as they turned out to be) gave astronomers a "much better chance than stars for the observation of gravitational lens effects." Both Einstein and Zwicky died before these extraordinary cosmic lenses were discovered. The first one was found, by accident, in 1979, and since then many more have been discovered. These lenses are created by the gravity of either massive galaxies or clusters of galaxies.

The Einstein Cross is one of the more celebrated examples of gravitational lensing. It, too, was discovered unexpectedly, in 1985, by astronomer John Huchra of the Harvard-Smithsonian Center for

Astrophysics, in Cambridge, Massachusetts. Huchra and colleague Margaret Geller and their team were measuring the velocities of eighteen thousand bright galaxies in the northern sky. The farthest galaxy in their study was about 700 million light-years away. One night, Huchra was observing with a 4.5-meter telescope on Mount Hopkins, about 30 miles south of Tucson, Arizona, when his technician, who was observing with a smaller telescope, called. He had found a galaxy with a spectrum that did not make sense. Huchra observed it with a better instrument and immediately recognized it as a quasar. But there was a problem. Quasars, which are the universe's brightest objects, are from a very distant past; they lie far beyond the regions that Huchra's team was studying and shouldn't have shown up in the observations. Sure enough, when Huchra examined the spectrum and its redshift, he realized that the quasar was nearly 8 billion light-years away. Something very strange was going on.

There was yet another mystery surrounding the unusual object. When Huchra studied the spectrum at the center of the object, he detected a galaxy that was only about 500 million light-years away. Whatever they had found seemed to be a mixture of a nearby galaxy and an astonishingly distant quasar. More observations followed, and the astronomers realized that they were probably staring at a gravitational lens.

Better telescopes eventually revealed more details. The nearby galaxy was warping spacetime and bending the light from the distant quasar toward us. Theory predicts that if the foreground galaxy and the object behind it are aligned with our line of sight, the background object should appear as a perfect ring. In Huchra's case, the elliptical nature of the foreground galaxy and its position relative to the quasar along our line of sight is such that the gravitational lens creates four images of the quasar instead of a ring, hence the name "Einstein Cross." It was another spectacular confirmation of a celebrated theory, adding to general relativity's status as the bedrock of cosmology.

By the 1990s, various elements were coming together in our understanding of how the universe began and how it was evolving. It

was becoming possible to build computer models using general relativity, based on the idea that the universe is homogeneous—meaning
that at very large scales it has more or less the same structure wherever you look. Dark matter—specifically, cold dark matter—was
incorporated into these models, and the models were successful at
showing structure formation on the scale of clusters of galaxies and
clusters of clusters. But there were a number of hints that the universe contained something more than just normal matter and dark
matter.

In the 1990s, two separate teams of astronomers had trained their
telescopes to look deep into the nearby universe. One was led by
Saul Perlmutter of the Lawrence Berkeley National Laboratory and
the other by Brian Schmidt of Australian National University and
Adam Riess, who was then at UC Berkeley. They were looking at individual stars in galaxies hundreds of millions, even billions, of light-
years away. Normally, it's impossible to see individual stars so far
away. But these astronomers were after stars that were in their death
throes—the exploding stars known as supernovae. Supernovae, in
their final moments, can outshine their host galaxies and act as beacons, lighting our way to distant parts of our universe. The astronomers were hoping to confirm that the expansion of the universe was
slowing down with age.

Lidman had been a member of Perlmutter's team. Both teams
had been monitoring a special type of exploding star known as a
type-Ia supernova. Such stellar explosions have become astronomy's
"standard candle," meaning that we can determine, within limits,
their absolute—that is, intrinsic—brightness and, as a result, their
distance from us. This calculation is based on the relationship between how long such a supernova takes to reach peak brightness and
then wane. This period can be accurately measured and is strongly
correlated with the supernova's absolute brightness. Then astronomers measure its apparent brightness—its brightness as seen from
Earth—and its redshift. The apparent brightness, when compared
with the absolute brightness, tells us the distance to the supernova,

and its redshift is a measure of how much the universe has expanded since the supernova detonated. The idea was to gather data both from nearby and distant supernovae and then compare them to see how the expansion of the universe had changed over time.

When the teams began their work, there was a consensus among astronomers that the expansion of the universe would have slowed over time, as gravity started to exert its influence. But it was an uneasy consensus. By the 1990s, astronomers were fairly certain of at least one big unknown in the composition of the universe — cold dark matter, the unseen mass of galaxies. Still, said Lidman, "there had been hints over the years that there had to be something in addition to ordinary matter and cold dark matter in the universe."

These hints came from two independent investigations. One had to do with the age of the universe. Cosmological models that incorporated both ordinary matter and cold dark matter were coming up with an age that was less than the age of some globular star clusters. Globular clusters are tightly bound systems of up to a few million stars, and many such clusters are thought to have formed just a few hundred million years after the big bang; if they were older than the universe itself, something was seriously wrong. Another hint came from studies of the numbers of galaxies. As you go back in time, or toward fainter and fainter galaxies, you should see fewer and fewer — because in a younger universe galaxies are still forming and are not as numerous as in an older universe. Or at least that's what models containing only ordinary matter and cold dark matter were telling cosmologists. But astronomers were finding something very odd: As they looked for fainter and fainter galaxies, deeper back in time, the number counts were increasing, and very, very quickly. "If you are starting to see many more galaxies than you expected, then either the galaxies are becoming more and more in number, or they are becoming brighter, or the volume of space is increasing faster than you expected," said Lidman. The first two reasons were ruled out, and the third was too weird to contemplate.

With these cosmological clouds hanging over them, the supernova teams began their survey in the mid-1990s. And what they found stunned the world of physics: The expanding universe, instead of slowing down with age, or even just coasting along, is actually speeding up.

Fortunately for cosmologists, Einstein had given them a clue as to how this might happen. After he had finished developing his general theory of relativity, he was left with one uneasy prediction: The universe could not be in a steady state; it had to contract under the influence of gravity. Perturbed by this, Einstein uncharacteristically fudged—he added what he called a "cosmological constant" to his equations, which allowed him to rein in a collapsing universe and keep it in a steady state. This was before we knew there were other galaxies, let alone that they were rushing away from one another. Einstein was quick to retract the cosmological constant once Hubble's observations contradicted it. Oddly, what he did, in effect, was provide a mechanism to explain how the universe could expand: His cosmological constant assumes that space has an inherent energy and that this energy counters gravity. The supernova studies indicate that something akin to this is indeed happening. Some energy is countering gravity and causing the universe's expansion to accelerate. Along with dark matter, there is now another puzzle: dark energy (sometimes referred to as the cosmological constant). Together they form the bulk of the universe.

"It is pretty hopeless, isn't it?" said Lidman. "We don't really understand much of the universe, dark matter is twenty-odd percent, dark energy is seventy-odd percent, so ninety-odd percent of the universe is a complete mystery."

The four massive telescopes at Paranal are playing a big part in unraveling this mystery. Two aspects of the VLT are crucial: a state-of-the-art spectrograph and the enormous 8.2-meter mirrors. But before these giant telescopes swing into action, smaller 4-meter-class telescopes in Chile and Hawaii, among others, scan the skies. Smaller-aperture telescopes have a bigger field of view and so can look at larger sections of the sky than 8-meter-class and 10-meter-class telescopes can. These smaller scopes take snapshots of the sky on dark, moonless nights and then look at the same parts of the sky again about three weeks later. Astronomers compare the two images of the same section of sky. If a supernova has exploded in the intervening weeks, it shows up in the second image. "Typically we catch most of our supernovae when they are at their peak brightness, or just a little bit before that," said Lidman. "The surveys have been designed so that the probability is that you will catch them as they are coming up."

At this stage, the astronomers are not sure what kind of supernova they have found, and that's where the VLT comes in. Its 8.2-meter mirrors are big enough to gather light from supernovae that exploded when the universe was just half as old as it is today. This light then passes through an advanced spectrograph, a far cry from the one Slipher used to determine the redshift of the Sombrero galaxy. The Brashear spectrograph that Slipher worked with weighed well over 400 pounds and this included counterweights that had to be added to the 24-inch telescope so that it could accurately track an object in the sky. A hundred years later, the twin spectrographs used at the VLT (called FORS1 and FORS2) weigh more than 2 tons each and hang beneath the bowl-shaped truss that supports the primary mirror. And while the Brashear spectrograph split light into its constituent colors to be recorded on photographic film, today's spectrographs use large CCD (Charge-Coupled Device) arrays — souped-up versions of the kind found in digital cameras. Far better than the human eye, they are designed to be sensitive to the redder wavelengths, since the more distant the object being observed, the greater the redshift. These spectrographs help identify type-Ia supernovae, which, for example, have a telltale silicon spectral line that other types don't.

The effort now is to see farther and better. The larger the pool of type-Ia supernovae, the easier it will be for astronomers to characterize the rate of expansion and the nature of dark energy. Does the density of dark energy change with time, or does it remain constant? The best candidate theory to explain dark energy remains Einstein's cosmological constant (which posits a dark-energy density that stays the same over time), but so little is known about this mysterious aspect of spacetime that other theories, some quite exotic, remain in the running.

"These projects are going to last a few years," said Lidman. "They need to build up the numbers. Previous results were based on tens of supernovae. The next generation of experiments is hundreds of supernovae—they are an order of magnitude larger, and probably an order of magnitude better in quality as well."

And the race is on, in a manner of speaking, to collect the best data possible. Perlmutter's Supernova Cosmology Project and Adam Riess's rival Hubble Higher-Z Supernova Search (the "Z" stands for redshift; the higher the z-number, the farther away an object is) are two high-profile projects.

"There is competition, to be sure," said Lidman. "How well you observe, how high the redshift is."

"Is redshift a macho thing among astronomers?" I asked.

"Of course it is," said Lidman, laughing.

"Just like the size of the telescope?"

"Yes," said Lidman. "Because you are pushing back the barriers. You are going back a little bit further in time. It may seem macho, but it is also important."

It was my last night in Paranal. I had become accustomed to the overwhelming darkness at night. Valentina and I had taken a walk the previous evening. There was no moon, and the only thing that relieved the blackness was the bejeweled sky. As my eyes adjusted, the stars increased in number, until the sky seemed to spill over with them. The disk of the Milky Way was a band so thick I could barely make out individual stars. But despite its stellar profusion, the Milky Way has many dark patches. These are created by clouds of dust, which obscure even the bright light pouring out from the center of our galaxy. So dramatic are these dark regions that those who have lived under these southern skies have seen creatures in the patterns created by starlight and dust; the Incas saw them, and the most famous example is the Emu, seen by the aboriginal people of Australia.

But as magical as the night sky was, a strange claustrophobia had enveloped me. Maybe it was the room I was sleeping in. Initially,

I had admired its sparse and utilitarian furnishings, but tonight it seemed bleak. The blinds had to be pulled shut each night, making the room seem even smaller. I started to understand why some astronomers compared being in Paranal to living in a submarine. You are hemmed in, if not by the dome (luxurious though it is), then by the harsh desert outside.

Earlier in the day, Roberto Tamai, a rugged, rangy Italian engineer, had expressed similar sentiments. He had left a cushy job building wind tunnels for the European Space Agency and relocated with his wife and two young daughters to Antofagasta, and he was now the chief of engineering for the Paranal Observatory. "I don't know if others have this feeling, but after a [tour of duty] here, you start to hunger for green, for blue, for something different," he said. "When I drive down to the city, it is absolutely special. It is difficult to explain to people who have not felt this. These are the feelings of many people who work in mines, or in the middle of the ocean on an oil rig."

That night, closeted in my room, I felt a new appreciation for those who made a home in this desert—especially one who had worked hard to turn Paranal into an oasis: Carlos Arriagada, the gardener. I often caught glimpses of him amid the thick foliage inside the dome, as he worked in blue overalls, workman's boots, wraparound safety glasses, and beige baseball cap. Thick brown-gray hair peeked out from under his cap. Behind the large glasses were kind-looking eyes, surrounded by generous wrinkles that accentuated the gentleness of his weathered face.

A few years earlier, Carlos had arrived in Paranal a depressed man. He had endured a difficult separation from his wife, had lost his job, and had been living on his sister's charity. The garden gave him a lifeline. Carlos spoke at length about it, and at one point tears welled up in his eyes. He made no attempt to hide them. "This is the culmination of my life," he said. "I'm going to devote the rest of my life to this. I'm in love with this garden, and in love with Paranal. The astronomers are looking for the unknown in the universe. I'm devoted to the things I know, and we are a very good complement."

But unknowns have their appeal, and nothing is tugging harder at

cosmologists than the mystery of dark energy. The supernova studies are turning out to be just one arrow in their quiver. Other experiments are being designed to decipher dark energy, and one of the
more remarkable involves counting the galaxies in the universe bit
by bit. The evolution of galaxies and the change in their numbers
over time is intimately linked to the nature of dark energy. Only a
very special set of instruments can do this: an iconoclastic 10-meter-
class telescope paired with a rumbling giant of a spectrograph. Together, they sit atop a sacred volcano in Hawaii, nearly 5,000 feet
higher than Cerro Paranal. That's where I was headed.

5 · FIRE,
ROCK, AND ICE

FOR SEVERAL DAYS, Mauna Kea—Hawaii's highest and most sacred mountain—had been masked by low-lying clouds. The day the clouds disappeared, I was driving along Highway 19, which snakes along the coast where the Pacific meets the northeastern flank of Mauna Kea. The lush landscape is shocking in its abundance. The road can barely contain the vegetation. Even the roadcuts, which must once have been bare when freshly hewn to build the highway, are now covered by a profusion of grass, shrubs, and trees. Isabella Bird, a Victorian-era English traveler, journeyed along the same coast in 1872. She was on horseback at the time, riding up and down the steep gulches that Highway 19 spans so effortlessly today. Bird found the gulches irresistible and said they had a "distracting beauty." These narrow and deep gorges have been carved by the unlikely combination of lava flows, snowmelt, and tropical rainwater. Today, from the road, you can occasionally glimpse the white of a waterfall plunging down a gulch through the dense green that smothers the slopes. Bird wrote: "The cascades are most truly beau-

tiful . . . falling into deep limpid basins, festooned and overhung with
the richest and greenest vegetation of this prolific climate, from the
huge-leaved banana and shining breadfruit to the most feathery of
ferns and lycopodiums."

The landscape on Hawaii, the Big Island, changes with astonishing rapidity. Soon I left the coast and reached Waimea, a region more
reminiscent of pastoral England than the tropics. It was there that I
glanced up and saw the top of Mauna Kea for the first time. Wispy
clouds still swirled around the summit, but the strong winds were
making light of them. Its flank exposed all the way from Waimea to
the summit, Mauna Kea stood solid and mighty. Hawaiians call it
Mauna Kea Kuahiwi Ku ha'o ika malie, "the astonishing mountain that
stands in the calm." Atop the summit, the silvery white domes of
some of the world's finest telescopes glinted in the light of the setting sun, looking out not just to the sea but beyond, to the very beginning of time.

The telescopes were beautiful to me, but to many Hawaiians who
grew up with the pristine profile of Mauna Kea, the domes are ugly,
and painful to see. In some traditional Hawaiian accounts, Mauna
Kea is regarded as the firstborn of the island of Hawaii and the island's people as descended from it. Ancestors are one's connection to
the past, spiritual and otherwise, and Mauna Kea is the most sacred
of ancestors. No wonder, then, that when the giant domes of the 10-
meter-class telescopes appeared on the summit in the early 1990s,
those who had always looked to the mountain for spiritual guidance
felt violated. Their mountain had been desecrated.

James Kealii Pihana, a grizzled sixty-five-year-old U.S. Army veteran and now a Mauna Kea ranger and cultural caretaker, recalled
the time when he was summoned by the ruling chiefs of Hawaii
to address the clash of cultures: astronomy on one side and deeply
wounded sensibilities on the other. Pihana belongs to a royal blood-
line himself, and priesthood courses through his veins. Still, he had
to reeducate himself about Mauna Kea, its cultural and spiritual significance, to gauge the extent of people's anger.

"As I was getting to learn about this particular mountain, it sort of hurt here in the heart," he said, smacking his chest with a closed fist. We were sitting in a chalet-like residence for astronomers, a few thousand feet below the summit of Mauna Kea. "How dare a foreign culture come here and take over an area that is considered very sacred to our people, with disrespect." But Pihana's mind began to change the more he learned about astronomy. He realized that Mauna Kea was one of the world's best sites for viewing the skies. In the early 1960s, after a tsunami had devastated parts of the Big Island, Mitsuo Akiyama, of the Hawaii Island Chamber of Commerce, started looking for ways to boost the island's economy. He knew of the clear night skies above Mauna Kea, so he wrote letters to many U.S. and Japanese universities and research institutes, inviting them to come to Mauna Kea to establish an observatory. Only one man replied: the astronomer Gerard Kuiper (the Kuiper Belt—huge lumps of ice and "dwarf planets," such as Pluto, that orbit the sun just beyond Neptune—is named after him). When Kuiper heard from Akiyama, he was immediately interested. Further testing established without doubt that Mauna Kea, 13,796 feet above sea level, was indeed ideal for observing the night skies.

Just like Cerro Paranal in Chile, Mauna Kea benefits from an inversion layer of clouds that lie far below the summit. These clouds, normally about 2,000 feet thick, stay low, their tops reaching an altitude of only about 6,000 feet. Mauna Kea towers above them. The inversion layer keeps the moist, turbulent air from rising. And for thousands of miles around the mountain, there is only ocean; there are no mountain ranges to roil the atmosphere. These factors make the air around and above the summit of Mauna Kea dry and stable, perfect for astronomy. Given such favorable conditions, astronomers commissioned a 2.2-meter telescope in 1970. Meanwhile, sites in the Chilean Andes were also being identified and developed, so astronomers were assured of unhindered views of both the northern and southern skies. The entire cosmos had opened up to humankind.

The conditions on Mauna Kea proved to be so good that soon more telescopes sprouted on the mountain (there are now thirteen in all). And it was this rampant construction, without due regard to what the mountain meant to the native Hawaiians, that finally led to the conflict of the 1990s. Things got so ugly, said Pihana, that "it was almost going to shut this mountain down for astronomy. People were really angry." He helped pacify the Hawaiian community. "We can also be forgiving as a race," he said. "And this is where priesthood comes in, to keep the people calm." But bringing them around wasn't easy. "It took a while to hear both sides," Pihana remembered, "sitting in meetings, week after week after week."

Today there is relative peace. Each side acknowledges the importance of Mauna Kea to the other. Construction is now limited to a certain region of the summit. Sacred burial sites near and on the very top of the mountain are off-limits. As a steward of the mountain, Pihana is happy that his community can offer Mauna Kea to the world. "It is important that our people understand the importance of the role this mountain will play in the future," he told me, "not only of Hawaii, but the entire world, the human race."

Mauna Kea is now home to some of the most powerful telescopes on Earth, including the iconic optical twins: the 10-meter-class Keck I and Keck II telescopes, which broke from the tradition of telescope mirrors made from a single piece of glass and thereby smashed the 8-meter barrier. Their astonishing light-gathering ability, combined with a spectrograph that can simultaneously measure the velocities of more than a hundred galaxies with unprecedented precision, has made Mauna Kea a unique vantage point from which to observe not hundreds or thousands but tens of thousands of galaxies. Studying these galaxies in the depths of space is allowing cosmologists to understand how the expansion rate of the universe is changing over time—crucial if we are to crack the conundrum of dark energy. The DEIMOS/DEEP survey, headed by Marc Davis of the University of California at Berkeley and Sandra Faber of UC Santa Cruz, complements the studies being done by Saul Perlmutter and others. Davis and Faber are using galaxies instead of supernovae to trace the universe's expansion, thus providing an independent confirmation of the nature of dark energy. None of this would have been possible, however, if it hadn't been for a bold, confident, even cheeky physicist who dared to break the mold and design a telescope of limitless possibility.

As Mauna Kea was captivating astronomers in the late 1970s, a Berkeley physicist in his early thirties was imagining an audacious telescope that would one day become synonymous with the mountain. At the time, the University of California was debating its future in astronomy. Its Lick Observatory—on Mount Hamilton, near San Jose—had a 3-meter telescope, but with nearby Silicon Valley booming, the increasing light pollution was making that observatory less and less effective. The university astronomers started discussing whether they should get out of optical ground-based astronomy altogether or build a bigger telescope at some darker site, and they invited Jerry Nelson, a researcher at the Lawrence Berkeley National Laboratory, to join the committee that would decide UC's direction.

Some thirty years later, Jerry Nelson remembered this period well. We met at UC Santa Cruz, in his light-filled corner office, whose large windows looked out at a grove of redwoods. Well known for his taste in Hawaiian shirts, he didn't disappoint. Even on a very un-Hawaiian day in wet Santa Cruz, he wore a bright, patterned shirt, and throughout our chat he remained as cheery as his shirt, often breaking into a generous and toothy smile, which took up so much of his face that his eyes nearly disappeared.

His colleague Sandra Faber remembers a "brash young" Nelson at a meeting with David Saxon, then the UC system's president, to discuss telescopes. Nelson wanted to build the world's largest telescope and was asking for $5 million to start the project, a sum of money Faber thought was too low. "I feared Saxon would throw us out, because it was clear that we did not know how to estimate costs," she told me. "Fortunately, he didn't."

Nelson, for his part, was merely asking for money to start researching and developing the technologies needed for a bold new telescope. The astronomy world was awash with 3- and 4-meter-class telescopes, and the world's largest at the time was the 5-meter at Mount Palomar in California. Why keep creeping along this path?

thought Nelson. "I wanted to double the 5-meter," he told me. But building a 10-meter-class telescope with existing technology was courting disaster. The primary mirror would have to be made of a monolithic block of glass. At the time, astronomers had yet to conceive of the high-tech monolithic mirrors for the 8-meter-class VLT at Paranal in Chile, let alone something bigger. Building a 10-meter primary mirror from a single block of glass—high-tech or otherwise—was madness: Polishing it would be no small task, and even if that could be done, the mirror's weight would be crushing. "To build a monolithic telescope is a cruel game," said Nelson. "We needed to break the paradigm, so I thought, 'Why don't we segment the mirror?'"

No one had ever thought of building an optical telescope mirror out of smaller segments. The wavelength of visible light collected by optical telescopes ranges from around 400 to 700 nanometers, so the imperfections in the curvature of the mirror need to be considerably smaller—of the order of a few nanometers—otherwise they lead to blurry images. Segmenting would likely introduce curvature variations large enough to doom the mirror. Nelson, however, was confident it could be done. "I have the arrogance that physicists tend to have," he said. "Everything is understandable. I learned the laws of physics—I just need to apply them to this situation." So he applied them to the problem of assembling small mirror segments to create a smooth mirror 10 meters across.

There are limited choices when it comes to creating a mosaic-like, or tessellated, surface with regular polygons. You can pick from one of three: equilateral triangles, squares, or hexagons. Other polygons will lead to an irregular tessellation, which complicates the mathematics needed to control them. Nelson settled on hexagons, because the blanks, to be made from the glass-ceramic Zerodur, came in circular shapes, and cutting a hexagon would be the simplest and least wasteful. Thirty-six hexagonal mirrors, each 1.8 meters across, would be needed to create a 10-meter primary mirror. To put this into context, each segment would be nearly three-quarters the diameter of the 100-inch Mount Wilson telescope.

That was just the first step. Next, the thirty-six segments would have to be aligned with nanometric precision to create the perfect surface; the slightest misalignment would render the telescope worthless. Nelson designed what he calls edge sensors, which detect the tiniest of movements between two segments. When such movements are identified, 108 actuators, or pistons — three for each segment, push or pull at the segments, keeping all thirty-six mirrors aligned. The alignment is done twice each second. Nothing like this had ever been attempted before.

However, none of it was intuitive, and Nelson had first to convince himself that the mathematics behind his design was sound. In fact, the entire project left many astronomers shaking their heads. Nelson remembered a meeting in Tucson at which he had to try to convince Frank Low, one of the pioneers of infrared astronomy, that the segmented mirror would work as designed. He couldn't. "And [Low] was a smart guy," said Nelson. "When you disagreed with him, you had to be pretty confident, because he knew a ton. But he was working off his stomach, he hadn't thought it through. He just thought he had superb intuition. But this was a new problem, so his intuition wasn't so superb in this new environment."

Other astronomers fretted about the 3-millimeter gaps between segments. The total area lost to the gaps is about seven-tenths of 1 percent of the total light-collecting area of the mirror. That was insignificant as far as Nelson was concerned, but many astronomers did not think so. "Some would say, 'Oh, those gaps are going to kill you,'" Nelson recalled. They worried that the gaps would reduce the telescope's ability to collect enough light, making it harder to see faint galaxies at high redshifts.

Nelson and his colleagues listened to such complaints attentively, for they were out on a limb with the new design. "We worked really hard at understanding our critics," Nelson told me. "You have to be really humble and really pay attention, even if most of the criticism is wrong." Eventually, Nelson was convinced that the design would work — the mathematics showed beyond a doubt that the gaps were not big enough to matter. There was, however, one more major hur-

dle. The surface of the planned 10-meter mirror would follow more
or less the curve of a parabola, not a sphere—thus each segment
would be parabolic, or aspherical. But polishing a mirror to take on
an aspherical surface is not easy—never mind thirty-six of them,
each with a slightly different curvature. This is because when polishing, the round-headed tool and the flat piece of glass need to be in
constant contact to get a perfectly smooth surface—and that auto-
matically creates a surface with a spherical curvature.

To solve this problem, Nelson's team came up with a new technique that drew on the older one. They started with a glass blank
for each segment, loaded it with twenty-four weights at the periphery, and adjusted the weights to elastically deform the blank into a
very precise shape. Once the blank was deformed, opticians ground
a spherical surface into it, a relatively easy process. Then when the
weights were removed, the glass eased back into the required aspherical shape. The trick lay in loading the twenty-four weights precisely
for each segment. There were a few other niggles, but for the most
part the opticians had come up with a new way of polishing mirrors.

"By the mid-1980s, UC said, 'Well, this is all very interesting,
but where is the money going to come from?'" said Nelson. UC estimated that $70 million—an enormous sum at the time—would be
needed to build one 10-meter telescope on Mauna Kea, a site chosen
after studies showed that it offered the best seeing in the Northern
Hemisphere.

The effort to build the 10-meter segmented telescope attracted
considerable media attention in the San Francisco Bay Area, and the
coverage was noticed by a certain Edward Kain. On August 23, 1983,
Kain called Joe Calmes at the Lick Observatory and suggested his sister Marion Hoffman, who lived in Los Angeles, as a possible donor.
Hoffman was a widow. Her husband, Max Hoffman, an Austrian-
born businessman, had made a fortune in the 1950s selling foreign-
made cars in the United States. Often he was the only man between
the manufacturers in Europe and the dealers in the United States, introducing Americans to cars made by Mercedes, Porsche, and Volks-
wagen, among others.

UC approached Marion Hoffman, who soon became convinced that the project was exactly the sort of thing her husband would have wanted to fund. The Hoffman Foundation agreed to give $36 million toward the telescope, with the rest to be raised by UC and the telescope to be named after Max Hoffman. On December 15, 1983, Marion Hoffman shook hands on it with then UC president David Gardner. The very next day, Mrs. Hoffman, who had been suffering from throat cancer for some time, died on the operating table, without having signed anything.

Meanwhile, for the remainder of the money, UC had partnered with the California Institute of Technology, and Caltech found a benefactor in Howard Keck, the chairman of the W. M. Keck Foundation, which was funded by hundreds of millions of dollars from the sale of his Superior Oil Company. An arrangement was worked out in which two 10-meter telescopes would be built on Mauna Kea, one named for Keck and the other for Hoffman. "We were on cloud nine for a couple of weeks until Marion Hoffman died without finalizing the deal," Faber told me.

The remaining two trustees of the Hoffman Foundation—Marion Hoffman's sister and her secretary, who strongly disliked each other—reluctantly gave the University of California $36 million in stocks, bonds, and Renoir paintings, which UC promptly auctioned off. But when the two trustees learned of the deal with Howard Keck, they were incensed that the memorial to Max Hoffman would not be exclusive. The foundation's lawyers, egged on by Marion Hoffman's sister, tried to get the money back. "UC finally decided they didn't want to accept a gift that wasn't gladly given, so they gave it back," said Nelson.

Howard Keck decided to fund both telescopes himself, and so the two most powerful telescopes atop Mauna Kea—jointly operated by UC and Caltech—came to be named Keck I and II. It's these two telescopes, their domes a mere 100 yards apart, that give the summit its new, albeit controversial, profile when viewed from Waimea, 11,000 feet below.

In November 1990, Keck I saw first light, but it wasn't your typical moment of clarity. Nelson's team had put in place nine out of the thirty-six segments for Keck I, and already the telescope was as large as the biggest telescope at the time, the 5-meter on Mount Palomar. But they were having serious problems with the computers that aligned the segments. So the unveiling was an unusual event. The nine segments were not perfectly aligned—each was acting as a standalone mirror. Also, the computers were not yet able to get the telescope to track a star as it moved across the sky, so in the very first images a star showed up as a trail instead of a point—and each of the nine slightly misaligned segments created its own star trail. Had the segments acted in concert, there would have been just one trail, and had the telescope tracked the star, there would have been just one point. But to Nelson's eyes the nine star trails were beautiful—they were all exquisitely thin. It was clear to Nelson that all the segments were of equally good optical quality and were focusing accurately. "That's the kind of first light only a designer could love," he recalled with a laugh.

By 1996, Keck II was also ready for science, and the Keck twins soon became the byword for telescopes on Mauna Kea. They were capable of seeing farther than any telescope on Earth. But a telescope is only as good as the instruments that analyze the light it collects. What the Kecks needed was a spectrograph worthy of them. A reluctant Sandra Faber soon found herself drawn into building an instrument that was just as audacious as the two new telescopes.

A spectrograph, conceptually, is a simple instrument. It essentially hasn't changed since the days of the Brashear spectrograph that Vesto Slipher used to study the spectra and redshift of spiral nebulae. But the scope of the instrument had to be redefined for the 10-meter Keck telescopes. Analyzing the light collected by such a gigantic telescope is not all that easy to do. The light beam from a Keck is so big that the lenses for the spectrograph had to be correspondingly huge and heavy. Also, keeping pace with the telescope as it tracked a star or galaxy was going to be far from trivial. Undaunted, astronomers at UC Santa Cruz started designing DEIMOS, the DEep Imaging Multi-Object Spectrograph, a 12-foot-high, 20-foot-long, 8.6-ton monster that would move in and out of the Keck focal plane on rails.

To track an object in the sky, the Keck telescopes—which are altitude-azimuth telescopes—move in two ways. One motion, which controls the altitude, moves the telescope from the horizon, which is the lowest point of the visible sky, to the zenith and down again to the horizon on the other side. At every altitude, the telescope can swivel around its vertical axis. The amount it swivels is the azimuth angle, with due north being 0°, east being 90°, and so on. A combination of these two motions helps the telescope track any star or galaxy. But it also causes the image being formed by the telescope to rotate. In order to lock in on the image, the heavy DEIMOS, planned for use on Keck II, also had to turn on its axis, like a beast rotating on a skewer.

That, however, was not the only challenge. Imagine a star that's being tracked from the moment it rises in the east to when it sets in the west. The telescope starts at the horizon in the east and follows the star almost all the way to the zenith. But just before the star reaches the zenith, the Keck swivels around, all 300 tons of it, so that it can point the other way and begin its descent to the horizon in the west. This has to be done as swiftly as possible, to minimize the telescope's blind spot in the sky. "While the telescope is whipping around, the spectrograph has to rotate a hundred and eighty degrees also," said Faber. "That is a sight to behold, believe me. When you are done rotating, you have to pick up at exactly the right place, and you have to be aligned spot-on to pick up that object. You have to be perfect."

When you are looking at not just one galaxy but nearly 140, as DEIMOS does, the dance has to be even more precise. It's this ability to observe so many galaxies at once that makes DEIMOS special. It uses a 28-inch-long slitmask, an aluminum sheet with precisely machined openings, each aligned to a known galaxy in the patch of sky being observed. Only the light from the galaxies being observed gets through the slitmask; everything else is blocked. Each time a different patch of sky is to be observed, the slitmask has to be changed. Over the years, the staff at Keck has been kept busy machining nearly five thousand different slitmasks for DEIMOS, making it possible to survey tens of thousands of galaxies; a single-slit spectrograph would never come close to achieving the same results. DEIMOS became the undisputed star among spectrographs and remains the most advanced instrument of its kind.

Sandra Faber is the principal investigator (PI), or lead scientist, for DEIMOS. Building it was "the hardest thing I ever did in my whole life," she recalls—and Faber has done many difficult things. She is one of the few astronomers who has successfully straddled the divide that separates the users and the builders of telescopes. The first part of her career, starting with her graduate work at Harvard, established her credentials as an observing astronomer. For her thesis, she defined the relationship between the spectrum of elliptical galaxies and their sizes. She then moved to California to join the Lick Observatory, whose headquarters are at UC Santa Cruz. It was at Lick, in the mid-1970s, that Faber did her best-known observational work, with her student Robert Jackson, this time finding the link between the brightness of elliptical galaxies and the spread of the velocities of stars in them. It came to be called the Faber-Jackson relation.

By the mid-1980s, Faber was being pulled to the other side. She got involved in the construction of the Keck telescopes and the Wide-Field Planetary Camera on the Hubble Space Telescope, honing her skills on the design of razor-sharp optics. The experience would serve her well. In 1992, UC Santa Cruz astronomers Garth Illingworth and David Koo started talking about building a multiobject spectrograph for Keck II. Faber got involved. "I went into the [DEIMOS] project thinking (a) I'm not in charge, and (b) I am surrounded by skilled and knowledgeable people," she remembered. But these comforting notions were soon dispelled. The others left one by one, and Faber "emerged as the tent pole, the standard bearer," in charge of the DEIMOS team.

I visited Faber not long ago at the hilly, wooded campus of UC Santa Cruz. The path to her office cut through groves of redwoods and lichen- and moss-coated trees, all of which spoke of life and longevity. In her sixties by now, her blond hair cropped short, Faber looked slighter than she did in photos, no doubt the aftereffect of recent hip-replacement surgery. She talked about DEIMOS as eagerly as a mother talks about her child. DEIMOS had consumed her life for a decade, some of the work done while she lay prone, unable to move because of debilitating back pain. She has lived with back pain since 1970, when she hurt herself while using the 36-inch telescope at the Kitt Peak National Observatory in Arizona. As a graduate student, she had been allotted five precious nights for observation. Faber fell off the observing platform on the very first night. She landed on a concrete floor five feet below and hit her head on it. Always diligent about keeping a log, her notes for the night end with: ". . . fell off platform and closed for night (1 A.M.)." After treatment for a mild concussion, Faber continued observing for the next four nights, occasionally groaning in pain as she moved.

Her chronic back problems certainly made DEIMOS a struggle. And there were other challenges. For starters, some astronomers doubted the need for a multi-object spectrograph. Faber recalled a particularly testy exchange during a meeting of the funding committee. "I will never forget a statement by an astronomer at Caltech who was sitting on this committee and said, 'Well, I never want to look at more than one object at a time.' He shall remain unnamed, but it was a he." Now, more than fifteen years later, Faber was still fuming. "To have somebody tell me that there is no point in building a multi-object spectrograph in the first place and, given that you are going to build it, there's no point in building it twice as big, because two times something worthless is still worthless . . . ," she said, unable to quite finish the sentence. Faber's team prevailed, if not entirely. They got to build DEIMOS at only half its planned capacity—that is, it can observe only half as many galaxies in one go as originally planned. "[It] bugs me," added Faber. "I'll go to my grave feeling that way."

But even at half the capacity, DEIMOS broke through many barriers. One of its most innovative features is the camera. The light from Keck II passes through a slitmask and thence to a grating. Like a prism, the grating breaks up the incident light into its constituent colors and sends it to a camera. Those who work with commercial cameras know how difficult it is to get wide-angle images that are equally good at all colors. Camera manufacturers are always compromising, and the fewer the compromises the higher the cost. DEIMOS could not afford any compromises. To further complicate matters, it also had a very wide field of view for a spectrograph — 17 arc minutes of the sky, which is half as wide as the full moon. "It was a hard job," Faber told me. "You were trying to achieve these perfect images over this wide field of view, in different colors. This spectrograph had to work all the way from four thousand to eleven thousand angstroms, which is a very big color range." This is where Harland Epps proved invaluable.

Epps was the resident optics expert at UC Santa Cruz. He had helped design lenses for just about every major astronomical instrument, including the Hubble Space Telescope. To ensure consistency over DEIMOS's wide field of view and color range, Epps designed a behemoth: a complex, 600-pound camera with nine lenses, the largest of them about 13 inches in diameter. The design had twice the field of view of any camera he had worked on before. It was a stunner. But Epps didn't realize just how much work he had created for the opticians who had to grind and polish the lenses.

The lenses had steeply curved surfaces, many of them aspherical, making them a challenge to polish. And three of the lenses had to be made of calcium fluoride. Lenses made of glass focus light of varying wavelengths slightly differently, which was unacceptable for DEIMOS. The only known material that could negate this effect when mated to glass was calcium fluoride. The calcium fluoride lenses were the largest that David Hilyard, of UCSC's optics workshop, had ever worked on. Calcium fluoride's extreme sensitivity to temperature meant that even warm breath could crack them. "They can go away in a heartbeat," said Hilyard. One such lens cracked in the time it took him to go and drink a cup of coffee.

There was one more serious challenge to overcome. The mechanical movement of the constantly rotating DEIMOS causes the optical path to flex, shifting the wavelength of light falling on any given pixel of the camera's detector. Each pixel in the 67-million-pixel detector can be slightly more or less sensitive than its neighbors, so ensuring that the same wavelength of light falls on each pixel throughout an observing run is crucial for image quality. "I think the final genius of DEIMOS was the realization that we had to hold the image, the wavelength, very constant on the detector," Faber says. DEIMOS is the only spectrograph with a "flexure compensation system," which constantly measures how much the optical train is flexing and corrects it, so that the same wavelength of light always lands on the same pixel—or, at most, moves a quarter of a pixel away.

On the night I visited the Keck II control room, which is not on the mountain but down in the less rigorous precincts of Waimea, Faber was putting DEIMOS through its paces. She and Renbin Yan, then a graduate student at UC Berkeley, were anxiously monitoring the unusually stormy weather over Mauna Kea. High cirrus clouds still dominated the sky. "There is a big clear patch to the north. That in some sense makes it worse. We call these sucker holes: You get going and a cloud comes right over," Faber said, a touch ruefully.

Faber and Yan took some images of the pre-twilight sky to calibrate the camera on the DEIMOS spectrograph. "That worked really great," said Faber to the observing assistant up on the summit of Mauna Kea, who appeared over a video link. Faber then asked the assistant to turn the telescope toward their first object for the night—a speck of sky in a region called the Extended Groth Strip, in the direction of the constellation Ursa Major. Faber wanted to gather spectra of galaxies in that part of the sky, galaxies that were more than halfway across the universe.

Meanwhile, DEIMOS's flexure compensation system was acting up. Faber talked about the spectrograph as if it were alive: "It's gotten completely deranged," she said. A few frantic moments later, everything was fixed, and DEIMOS was ready for use. The telescope, too, was pointing correctly. Faber set DEIMOS in motion with a few clicks of a computer mouse. Somewhere up on the summit, inside the Keck II dome, a slitmask moved into position. Once the telescope was locked and tracking, DEIMOS started moving with it. An image of the sky as seen by DEIMOS appeared on the computer screen in Waimea. Four known stars could be seen in square holes at the edges of the slitmask. These were the alignment stars, and they were slightly off-center. Yan sent commands to the telescope, fine-tuning its position, and checked again. The alignment was better, but not perfect. One more command. A less experienced team might have checked again to see if the stars were centered in their boxes, but Faber and Yan moved on to observing the galaxies. And sure enough, not just the four alignment stars but each one of the nearly 140 galaxies had lined up with their respective slits. Inside DEIMOS, a grating moved into place. The light from each galaxy was now being split into a spectrum. Three exposures followed. Taking three spectra meant that for each pixel they would have three readings. If any pixel were hit by a cosmic ray, making it brighter than usual, it would be discarded (a cosmic ray is unlikely to hit the same pixel twice in three exposures), and the average of the two dimmer readings would be used.

Keck II and DEIMOS make a powerful couple. Keck's size lets it gather more light from faint galaxies than any other telescope today, so astronomers can see deeper into the universe than ever before. Then DEIMOS kicks into action. It splits this light into precise spectra that allow for accurate measurements of redshift, and thus the velocities of galaxies. And all of it is made possible by "the astonishing mountain that stands in the calm."

Hawaii's Big Island is dominated by two volcanic mountains: Mauna Kea and Mauna Loa. The former pips the latter for the honor of being the highest mountain in the Pacific by a mere 35 meters. But Mauna Loa outranks Mauna Kea as the largest volcano on Earth. It rises 4 kilometers above sea level, a staggering 17 kilometers from the seafloor, which has in turn been pushed down by the mountain's weight. If the heft of these Hawaiian giants hadn't depressed the seafloor, the weight of the legends that rest on them might have. According to traditional accounts, Pele, the Hawaiian goddess of volcanoes, went searching for a home at a time when "the air was surrounded with spiritual beings and a thin veil divided the living from the dead, the natural from the supernatural." She found a home on Kilauea, an active volcano beside Mauna Loa.

Legend has it that Pele would send a white dog to warn people of an imminent eruption. According to records kept by the U.S. National Oceanic and Atmospheric Administration's weather observatory on Mauna Loa:

> A white dog was first noticed by the observatory staff during the latter part of 1959 about ½ km below the observatory . . . Attempts to befriend or capture this mysterious dog, no matter how persistent, failed. The dog for some reason would have nothing to do with the observatory staff. In December 1959, Kilauea Iki [one of the two active craters on Mauna Loa] erupted and the dog disappeared. The dog reappeared at the observatory several months later and again was spotted periodically for a month or so and then disappeared again. This pattern of appearances and disappearances continued until 1966. Since then, no one has seen this mysterious white dog.

Despite the benevolent white dog, Hawaiian tradition says that when Pele's molten body moves, "the land trembles and the sky is afire with a crimson glow . . . Those present whisper in awe, '*Ae 'aia la o Pele, there is Pele.*'" Little wonder that Hawaiians would rather not get in Pele's way. Particularly when she fights Poli'ahu, the goddess of snow. Poli'ahu resides on Mauna Kea, and she and Pele are said to have fought for control of the summit of Mauna Loa. This struggle was played out on the saddle-like region between the two towering mountains. Glaciers once dominated the summit of Mauna Kea, and lava flowed from Mauna Loa, causing fire and ice to meet on the saddle. Early Hawaiians avoided the area, letting it be the realm of the fractious goddesses.

Today, a road cuts right through the region. The appropriately named Saddle Road is the only way to get to Mauna Kea's summit. It takes you from the island's west coast to the east coast, with Mauna Loa to the south and Mauna Kea to the north. Hawaiians consider the road unlucky. Many rental car companies do not allow their two-wheel-drive cars on it—not for any superstitious reasons but simply because it takes too long to retrieve the cars that get stranded.

Saddle Road is an old one-lane army road that has been widened with half lanes of makeshift asphalt on either side. The edges are heavily pockmarked, even cratered. Drivers prefer the middle, moving onto the shoulder only to pass or to make room for oncoming traffic. It's not a ride for the queasy. The road sometimes swerves and curves even as it rises and falls, leading to blind turns at the very top of a climb. You have no idea what awaits you around each corner, especially on a cloud-laden day.

Driving on this 14-mile stretch of rather rough road, you become acutely aware of one of the key advantages of the 10-meter Keck telescopes. The segmented design of the 10-meter primary mirror meant that it could be transported in smaller packages. When the Gemini North telescope, built by an international consortium of seven countries (the U.S., the U.K., Canada, Chile, Australia, Brazil, and Argentina) was installed on Mauna Kea, the story was entirely different. In June 1998, Gemini's 8-meter monolithic mirror became the largest ever to be taken up to the summit. It was so big that at some points the road to the summit had to be regraded and road signs had to be temporarily removed. When the telescope's twin, Gemini South, was installed atop the nearly 9,000-foot-high Cerro Pachón in Chile, crews had to widen one of the road tunnels by as much as six feet. When it comes to building telescopes with monolithic mirrors, civil infrastructure has become as much of a limitation as the weight of the mirror, or its grinding and polishing.

Eventually, the nerve-wracking stretch of Saddle Road joins a newly built multilane federal highway. To the right looms the summit of Mauna Loa. Hundreds of years of lava flows have patterned its gentle slopes in shades of brown and black. To the left is Mauna Kea, its flanks scarred by deep gulches cut by ten-thousand-year-old glacial melt. As the road approaches the exit to Mauna Kea, the lava fields on either side of the road get bigger. Light-brown, jagged, jumbled-up stretches dominate—Hawaiians call it a'a lava, mimicking the sound of someone goose-stepping over the sharp rocks. Even more dramatic, and coming right up to the edge of the road, are rocks that look as if they were made of giant, black, twisted rope, the result of slow, ponderous lava flows. The landscape is also dotted with cinder cones — small conical hills of volcanic rock fragments. None is stranger than the one that marks the turnoff from Saddle Road toward Mauna Kea. The *Pu'u* (Hawaiian for "cinder cone") is lush green on the windward side and dry on the other, mirroring the schizophrenic beauty of this island.

The turnoff leads to the *Hale Pohaku,* Hawaiian for "stone house," the residence for astronomers. The residence, which is at an altitude of 9,000 feet, had actually been just a stone house at one time, hence the name. Today, scores of astronomers who work at the summit spend the night at the upgraded chalet-style building. It's also a place for a newbie to stay and adjust to the altitude before heading farther up.

The next four-mile stretch of road to the summit is unpaved and extremely steep, and its washboard surface makes for a rattling ride. Accidents are not uncommon on this steep section, especially on the descent, when a vehicle's brakes can overheat and fail. "Maybe once a year, we'll get a two-wheel-drive vehicle that tumbles and falls; they can't stop their car, so they end up going over the edge," said Laura Kinoshita, the Keck Observatory's public relations officer, who was driving our comfortable SUV. "We have had fatalities. Recently we had a Jeep Wrangler carrying six passengers. Two were killed when the Jeep went over." I clung a bit tighter to the door handles.

We climbed higher, leaving behind a thick bank of clouds — the inversion layer. Mauna Kea cast a shadow on the clouds. Finally we arrived at the summit, just moments after the sun had set. But even the dying sun seemed capable of setting alight everything that stretched before us. The lower bank of clouds looked as if fires raged beneath them. The bottom layers of high, storm-driven cirrus clouds burned bright. Between these two layers of clouds were the Keck domes, silhouetted against a sky replete with yellows, oranges, and reds.

Inside the Keck II dome, it was dark. There were no lights. The telescope wasn't being used yet for observing, but the dome was open. Despite the high cirrus clouds, a dozen or so bright stars were visible through the open dome. It was quiet inside, except for the cooling system of one of the telescope's instruments, its high-pitched pistons pumping refrigerants in long birdlike chirps. Once my eyes adjusted to the almost total darkness of the dome, the opening seemed to fill up with stars, perhaps a hundred of them, that hadn't been visible moments earlier. I looked down at the primary mirror and caught a glimpse of it through the tangle of trusses that held it in place, a flash of silver, a bright star reflecting off its shiny, polished surface. For a brief instant, Hawaii and its overwhelming assault on the senses disappeared. A quiet descended. It was just the dome, the telescope, and starlight that had traveled for thousands of years.

It was too dark to see DEIMOS, but it was there inside the dome, ready to rumble into action. Using the Keck and DEIMOS, night after night, year after year, astronomers have been observing tens of thousands of galaxies speeding away from us. They are members of the DEEP (for Deep Extragalactic Evolutionary Probe) team, led by Marc Davis of UC Berkeley and Faber, and they are turning these observations into a measure of the expansion rate of the universe and using it to solve one of the most profound problems in modern cosmology: the nature of dark energy.

On a late February afternoon, the UC Berkeley campus was abloom—cherry and plum blossoms had arrived early that year. As if to offset the bursts of white and pink, a student a capella group, gathered near an imposing gate with ornate metalwork behind which the trees bloomed, was soulfully belting out Elvis Presley's "Blue Moon" — *"You saw me standing alone, without a dream in my heart . . ."*

I knocked on Marc Davis's open door. He turned around slowly. His right arm lay limp on his lap. A cane was propped up against the desk. Davis had suffered a stroke in 2003. On his computer, photos of him and his family at a ski resort were flashing by as screen savers. An accomplished skier for forty years before his stroke, he was learning to ski again. He welcomed me in. "I'm completely dead here," he said, pointing to his right limbs. "The arm is shot, useless. The leg is ninety percent. It's sort of there. My mental functions are ninety-eight percent back."

He seemed sharp to me, fully able to cast his mind back to his days at Princeton in the early 1970s, when, as a student of James Peebles and his colleague David Wilkinson, he had embarked on an ambitious study of galaxies. The idea was to test some new theories about the formation of large structures in the universe, on the scale of clusters of galaxies. After his Ph.D., Davis moved to Harvard and teamed up with astronomers there to carry out what became known as the CfA Redshift Survey, named after the Harvard-Smithsonian Center for Astrophysics, where the work was being carried out. They looked deep into the northern sky and mapped out more than two thousand galaxies. No one had ever attempted anything on this scale before. "We saw the universe, and it looked so strange," said Davis. "It didn't look anything like our preconceptions." Rather than finding galaxies distributed uniformly throughout space, Davis and his colleagues saw galaxies clustered together and clusters congregated in superclusters, forming long filaments and giant walls. Most important, they found voluminous voids of empty space. The galaxies were like dots of light on huge empty bubbles.

Davis moved to UC Berkeley in 1981, still befuddled by the universe seen in the CfA survey. Nothing in their theories could explain the structures. But two big new ideas came to the rescue. One was the theory of cold dark matter, and the other was Alan Guth's theory of inflation. So, what would the universe look like if cold dark matter and inflation were added to the mix? Faber, Joel Primack, and colleagues took the first stab at answering the question and were able to show how different types of galaxies could arise from seeds of dark matter. Then, to simulate the large-scale structure of such a universe, Davis teamed up with physicists at Berkeley and Stanford. They were stunned by the simulations. "By god, they looked so good," said Davis. "Let me show you something. It's right here on the door."

He got up with great effort and walked slowly to the door without his cane. On it were three identical-looking computer printouts, each showing a pizza-slice view of the universe with Earth at the apex and galaxies dotted around vast regions of empty space. "One of them is the original CfA survey," said Davis. "And two of them are completely fake." By fake, he meant they were simulations. They looked just like the CfA survey. His voice betrayed the excitement he must have felt when he first saw the results two decades ago. "We had a tremendous success in that we could explain large-scale structure in the universe," he said. "We couldn't believe it, but the simulations worked. This was the mid-eighties. We became famous."

They would have become even more famous had they realized they were sitting on a gold mine of data. At the time, cosmologists were struggling to pin down the density of the universe—which meant knowing the density of matter and the density of any energy that might be present in the fabric of spacetime. Observations in the 1980s were suggesting that our universe was nearly flat, which meant that Omega—the ratio of the total density to the critical density needed to make the universe flat—was close to 1. In the CfA survey, which measured the mass of galaxies in a given volume of space, Davis had enough information to calculate Omega-matter (the contribution to Omega from matter alone). But the team made the mistake of assuming that spiral galaxies had the same mass as elliptical galaxies. Spirals would turn out to be about three to four times less massive than ellipticals. Had they known this at the time, they could have figured out that the Omega-matter made up only about 0.25. Clearly, something else made up the rest of Omega. One candidate for the "something else" was the cosmological constant cooked up by Einstein, the energy density of the vacuum of space. "Had we realized [it] at the time . . . our data was more consistent with a cosmological constant than anything else," said Davis. But no one, Davis included, trusted the data. If they had, they could have shown that nearly 75 percent of the universe is made of some mysterious form of dark energy. As we saw in the last chapter, the more precise supernovae studies of the late 1990s claimed this discovery.

Understanding the nature of dark energy has become the central question confronting cosmology today. Does the density of dark energy remain constant, or does it change with time? Cosmologists parameterize this property of dark energy with the letter w, its so-called equation of state. If w is equal to -1, then the density of dark energy remains constant over time—this is what's referred to as the cosmological constant. But w might be less than or greater than -1. These questions came to the forefront about the time that Davis, Faber, and others were getting ready for the DEEP survey, using data from the Keck II and DEIMOS. Davis, a postdoc of his named Jeffrey Newman, and a graduate student named Brian Gerke realized that they could use the fifty thousand galaxies being targeted by DEEP to determine w.

The idea is deceptively simple. Take a volume of deep space—say, a volume that lies between 5 billion and 7 billion light-years from us. Measure the number of gravitationally bound structures, such as clusters of galaxies, in that volume. Compare that number with what you find in a similar volume of space in the nearby universe. The comparison should tell you something about w. The repulsive effect of dark energy is thought to have kicked in when the universe was about half its current age of nearly 14 billion years. That's when the expansion of the universe started accelerating. Most of the large-scale structures in the universe should have formed before this process started, because the accelerated expansion of space does not allow for galaxies to coalesce and form large clusters. If our universe has a cosmological constant (w equal to -1), then the DEEP survey should see large clusters forming in the volume of space being monitored by Keck II and DEIMOS.

So, you might ask, if it's that simple, why isn't everyone doing such studies? Well, to start with, you need a telescope that can see faint galaxies that are more than halfway across the universe. No telescope is better than the Kecks at doing that. Then there is the art of the spectrograph. To figure out which galaxies are gravitationally bound together into clusters, you need to determine their velocities very accurately. The velocity of a galaxy can be measured from the redshift of its spectral lines. This is where DEIMOS excels. It has excellent spectral dispersion: the ability to separate out the various spectral lines from a galaxy. "Almost everybody who has worked in this game before has shitty redshifts," Davis told me. "They are not very accurate, because they are using crappy dispersion. Typically, when you measure the accuracy of redshifts, it is within a thousand kilometers per second. Turns out that DEIMOS measures things to within thirty kilometers per second. That's astounding." The DEIMOS/DEEP data seems to be pointing toward $w = -1$, a universe with a cosmological constant. Einstein may have stumbled upon the right answer after all: Spacetime might have an energy density that does not change with time. Still, the telescope and the instruments on Mauna Kea are working at the edge of their capabilities, and the DEIMOS/DEEP survey cannot answer the question to better than 15 percent accuracy.

Even if dark energy turns out to be Einstein's cosmological constant, there is still something that bugs Marc Davis. Why does dark energy have the value it does? Nothing in the laws of physics can explain that. If the value had been slightly lower or slightly higher, the universe would have evolved very differently—maybe without any stars and galaxies, and probably without any life. As noted in the prologue, lately there have been suggestions from theoretical physicists that our universe is just one of many universes. In this multiverse, each universe has a randomly selected value for such fundamentals as the cosmological constant; we just happen to be in a universe that is favorable to life. The anthropic principle—the idea that our universe has the properties it does because we are here to say so and that if it were any different, we wouldn't be around commenting on it—infuriates many physicists, including Davis. It smacks of defeatism, as if we were acknowledging that we could not explain the universe from first principles. It also appears unscientific. For how do you verify the multiverse? Moreover, the anthropic principle is a tautology. "I think this explanation is ridiculous. Anthropic principle . . . *bah*," said Davis. "I'm hoping they are wrong [about the multiverse] and that there is a better explanation."

Back in Waimea, Faber espoused exactly the opposite view when it came to making sense of a universe whose fundamental constants and physical laws seem fine-tuned for the existence of life. To Faber, there are only two rational approaches to answering the fine-tuning question. "One is that there is a God and that God made it that way," she said. But Faber, an ardent atheist, rejects that. "The only other approach that makes any sense is to argue that there really is an infinite, or a very big, ensemble of universes out there and we are in one. The point about the anthropic principle, that is really its force, is that you have to believe in the ensemble you don't see. And the reason I think this [approach] works is because it has worked historically several times in the past."

Faber offered an analogy. Imagine a Greek cosmologist two millennia ago, she said, who had just discovered that the Earth is round (in an era when everyone else believed that the Earth was flat and the center of the universe). The cosmologist measured the curvature of the Earth and found its diameter to be approximately 8,000 miles. "If you thought like today's physicists, then you would be trying to explain this number, eight thousand miles, from first principles," said Faber. "That's ridiculous. However, if you were an imaginative Greek thinking anthropically, you could have said, 'This is one example out of many.'" In other words, Earth's diameter is an accident—a likely accident, if there are hundreds and thousands of planets. The Greek cosmologist's colleagues would have argued that the anthropic principle was not scientific, but the cosmologist would have been exactly right, said Faber. "She could have stood up and defended herself against her colleagues by saying, 'This is the right way to go. I'm two thousand years ahead of you, in that I believe there are planets out there and you don't.'"

To those who argue that anthropic reasoning is not scientific, Faber had an answer. "I think it could be terrifically scientific," she said. "I don't think that a question is scientific today and not yesterday just because we have developed some technology in the meantime. A scientific question is eternally good; we just don't have the means to address it right now. But we may later."

It was getting close to midnight. In the Keck II control room, Faber and Yan wound up their observing run and Faber diligently wrote up the log for the night. DEIMOS had balked a bit. She made a note of that.

We walked out and stood for a few minutes under the night sky. Could there be other universes out there? Faber believes that there are. The cosmologies of religions seek to give humans a sense of place and purpose. For Faber, modern cosmology does the same. "I think we all have an affection for Earth. It is really a short leap from there to feel an affection for one's universe, especially if you think, as I do, that there are other universes out there that are probably hostile to the point of having different physics that wouldn't make bodies like ours," she said. "I take comfort in the fact that ours is a beautiful universe, that we belong here, and that we fit. This is our home."

Trying to tackle the anthropic principle is more than an exercise in science: It has philosophical implications. Is our universe unique, and if so, why? Are there other universes out there? How will we know? One way to begin answering the question is to study greater and greater volumes of our universe. If the DEIMOS/DEEP survey of fifty thousand galaxies has upped the ante, the proposed Square Kilometre Array (SKA), a gigantic radio telescope, will further raise the bar by sweeping the universe for a billion galaxies. The SKA will be searching for signs of the primordial sound waves that rang through the early universe. The imprint of these reverberations on the large-scale structure of the universe contains clues to the universe's curvature. Is the universe absolutely flat or just the slightest bit curved? The answer is crucial to verifying theories that predict a negative curvature, some of which suggest that our universe is one of 10^{500} different universes. To understand just what it will take to build the SKA, I traveled deep into the heart of South Africa, into an arid land called the Karoo.

6 · THREE THOUSAND EYES IN THE KAROO

JUST SOUTH OF the Kalahari Desert is a region of South Africa where no physical feature proclaims its past too grandly. You have to be a geologist to look beyond the unrelenting barrenness; this is a land so arid that sheep farmers need hectares of grazing land for a single animal. The Khoisan people of southern Africa call it the Karoo, "the land of thirst." A few hundred million years ago, it was anything but. The shards of shale underfoot speak to a time when this land was at the heart of the supercontinent Gondwana, much of which was covered by a vast ice sheet. As Africa broke off and drifted north, the ice gave way to inland seas and lakes, meandering rivers and lush vegetation, dinosaurs and mammalian ancestors. Then came unbridled volcanic activity. Molten magma from deep within the Earth forced itself through fissures and solidified into rocks on the surface. Eventually the land turned to desert. And much of this dramatic geological and paleontological history has been preserved in the thousands of meters of sediment in the Karoo.

James Bryce, a British politician, traveler, and historian, wrote in

1897 that the powerful charm of the Karoo lies in its "primeval solitude and silence."

> [There are] minds to which there is something specially solemn
> and impressive in the untouched and primitive simplicity of a
> country which stands now just as it came from the hands of
> the Creator. The self-sufficingness of nature, the insignificance
> of man, the mystery of a universe which does not exist, as our
> ancestors fondly thought, for the sake of man, but for other
> purposes hidden from us and for ever undiscoverable—these
> things are more fully realised and more deeply felt when one
> traverses a boundless wilderness which seems to have known
> no change since the remote ages when hill and plain and valley
> were moulded into the forms we see to-day. Feelings of this kind
> powerfully affect the mind of the traveller in South Africa. They
> affect him in the Karroo.

South Africans call it the timeless Karoo. It seems appropriate that this ancient land—which provides such a clear window to
Earth's past—has been tapped to look even further back in time. Billions of years, in fact. For if South Africa gets its wish, the largest radio telescope ever conceived will be built here, a telescope so large
that its thousands of antennas will add up to a combined surface area
of a square kilometer—hence its name, the Square Kilometre Array
(SKA). The Karoo and similar land in the Australian outback have
been selected as the only sites remote and empty enough to host the
telescope. It is a two-horse race that the South Africans are determined to win.

The real winner when the SKA gets built will be cosmology. Not
only will it study a universe invisible to optical telescopes but it will
do so with great efficiency and in great detail, allowing cosmologists
to observe a billion galaxies stretching out from the nearby universe
all the way to when it was just 4 billion years old. Such a survey will
dwarf even the most ambitious study of galaxies being carried out today. The SKA will achieve its target not by looking for light from galaxies but by searching for clouds of neutral hydrogen. Each atom of

this primordial gas can emit a radio wave. The SKA's sensitivity will let it listen to hydrogen's heartbeat, using it to locate the remnants of the massive amounts of gas that were the precursors to galaxies. Today these blobs of hydrogen act as tracers for their host galaxies, which are otherwise too faint to see. The study of the distribution of galaxies over most of the universe's life will illuminate the expansion of space. The secret of dark energy is hidden within this expansion history: This mysterious component of the universe would have dictated how the expansion rate has changed over time.

The nature of dark energy and the expansion of the universe are tied to the curvature of spacetime. Today's measurements show the curvature of the universe to be flat, but the limits of experimental error are such that it's hard to be certain if the universe is exactly flat or the tiniest bit curved. It will take a telescope of the size and sensitivity of the SKA to reduce the experimental errors to a level where cosmologists can be confident. Will the SKA point to theories that predict negative curvature—that is, a universe that is open rather than closed? (A flat universe has zero curvature and a closed universe, analogous in two dimensions to the surface of a sphere, has positive curvature.) Since some of those theories also argue for the presence of a multiverse, this could be a key finding in favor of it. Or will the SKA rule out such theories? The first steps toward answering these profound questions are being taken in the Karoo.

The Karoo makes up about 40 percent of South Africa's land but is home to only about 2 percent of the population. In some places there is less than one person per square kilometer, on average. There were even fewer before this harsh land became home to the northern European settlers known as Boers, who fled the Cape when the British occupied the region and abolished slavery in the early 1800s. The Boers settled and built farms in the Karoo, though the manner in which they set up the farms was atypical. A farmer would first pile up some stones to serve as a beacon. To garner territory, he would start off from the beacon, go as far and as fast as his horse could take him in half an hour, pile some more stones, then ride another half-hour at right angles to it, and so on until he had marked a rectangle of land as his own. The amount of land was limited only by the speed of the horse. Inevitable boundary disputes followed, as more accurate ways of measuring land became available. True, the region's extreme aridity requires large tracts of land for grazing, but the size of the farms in the Karoo today—often running into thousands, even tens of thousands, of hectares—is also testimony to the anarchic land grab of the 1800s.

Adrian Tiplady, the project scientist for South Africa's SKA bid, drove me to the farms that have been proposed as the best site for the core of the Square Kilometre Array. The gravel road we were on cut a straight gash through the land. Farmhouses were rarely visible, tucked away somewhere, their presence indicated by the occasional dirt road that disappeared into the nothingness. In an hour of driving, the only signs of habitation were windmills, their metal blades glinting in the sun. The windmills power mechanical pumps that draw out the scarce groundwater. A few sheep came into view every now and then. The shrubs they munch on give the region's lamb a distinctive aromatic taste, and Adrian couldn't wait to dig into some succulent Karoo lamb as soon as we stopped for the evening.

The deeper we drove into the Karoo, the drier it became—not surprising, for we were headed north, where the Karoo eventually gives way to the Kalahari Desert, near the Namibian border. We passed an occasional grove of spiky blue agave cactus, the source of South Africa's version of tequila. A brown hyrax darted across the road. A strange rabbit-size creature, the hyrax is supremely adapted to the dry Karoo, preventing water loss by minimizing its urine production; hyrax urine is viscous, like treacle. Occasionally baboons could be seen standing on the road ahead of us, only to scamper into the bushes as we approached. A lonesome narrow-gauge railway ran alongside the road. Here and there a town would appear, like a mirage, its white rooftops shimmering in the 86°F heat.

"It's a fantastic landscape for a radio telescope," said Adrian. "To the south are hills, mountains, which really protect the site from radio frequency interference. To the north there is space for [the telescope to expand]. It's emptiness, complete emptiness. But you can

still access the site reasonably easily. You don't have to fly there."
This was a subtle dig at the Australian competition, which has chosen a site deep within Western Australia, almost 400 miles from Perth, the closest major urban center.

To nonscientists, radio telescopes are not the most obvious of instruments, unlike optical telescopes, which have mirrors and lenses to gather light. After all, we don't see or hear radio waves. Radio telescopes weren't obvious to scientists either, which might explain why the invention of radio astronomy was serendipitous. It happened in 1931, when a twenty-six-year-old physicist named Karl Jansky built a radio antenna near his place of work, the Bell Telephone Laboratories in Holmdel, New Jersey, where Penzias and Wilson would someday discover the cosmic microwave background radiation (see chapter 4). Bell Labs wanted Jansky to analyze the sources of noise that could potentially interfere with intercontinental radio communications. By then, AT&T, owner of the labs, had successfully demonstrated an overseas radio link between the United States and Europe and wanted to expand the service. But it was a low-frequency link and did not have enough bandwidth for a worldwide service; the answer was to move to higher frequencies, or shortwave radio. The trouble was that the shortwave band was plagued by atmospheric noise, and it was Jansky's job to characterize this noise.

Jansky built a 100-foot-long antenna. The only good remaining picture of Jansky and his antenna makes him look like an airman from the Wright brothers' era, his trousers tucked into his socks; the telescope looks like the boxy frame of a flimsy biplane, sans propellers and fuselage. The antenna stands on four Ford Model-T tires, which moved around a circular base, allowing the contraption to rotate and receive signals from any direction. Jansky did identify something that could mess up radio communications: electrical discharge from thunderstorms, both local and distant. But he found another signal, too—a mysterious hiss, or static, which changed over the course of a day, peaking daily with surprising regularity. However, it didn't keep exact time with a twenty-four-hour day.

The signal followed what is called a sidereal day, with a period of twenty-three hours and fifty-six minutes. Only stars keep time to a sidereal day. So whatever this hiss was, its source was outside our solar system. Jansky persisted—a trait that has helped many a radio astronomer turn serendipity into discovery—and established that the signal was coming from the Milky Way, possibly the galactic center. He had discovered the first cosmic radio waves. The story made the front page of the *New York Times* on May 5, 1933. But Jansky's employers, satisfied that their engineering problem had been solved, cut the project short. Jansky himself did no further work on radio astronomy.

"Enter Grote Reber"—as the radio engineer turned astronomer himself later wrote about his role, in a whimsical paper called "A Play Entitled the Beginning of Radio Astronomy." As a young Chicagoan, Reber found himself "enchanted" with Jansky's discovery and proceeded to investigate it further by building, in the summer of 1937, a proper radio telescope—a parabolic antenna 31.5 feet across, with a focal length of about 20 feet. Optical astronomers had been using parabolic mirrors to focus light of different wavelengths to a single point, and Reber decided to do the same for radio waves. At the focus was a receiver that amplified the weak signals.

Reber installed the dish in a vacant lot next to his mother's house in Wheaton, Illinois, a Chicago suburb. Like Jansky, he used parts from a junked Ford Model-T truck to operate his telescope. ("Clearly, Henry Ford played an important part in the beginning of radio astronomy," he would later quip in his memoir.) The telescope was a curiosity in Wheaton. People gawked and stopped to take pictures. Some even rang Reber's doorbell to ask him about it. Small planes, at a time when "air navigation rules were very lax or non-existent," would buzz the telescope to get a good look. Eventually, the curiosity wore off, so much so that "after school, some of the larger children would use the telescope as climbing bars." Thankfully, they weren't able to clamber onto the surface of the dish.

Such intrusions notwithstanding, by the early 1940s Reber had confirmed Jansky's discovery of radio waves from the Milky Way, and radio astronomy moved on to become a serious discipline. The basics of a radio telescope were by now in place: a parabolic dish to focus incoming radio waves and a receiver at its focus to amplify the signal. Most modern radio telescopes are variations on this theme.

Reber's place in radio astronomy is captured rather poignantly in a graph that plots the sensitivity of a radio telescope against the year in which it was built. Drawn by astronomers designing the Square Kilometre Array, the graph starts at 1940, with Reber's telescope assigned a sensitivity of 1, the baseline against which all future telescopes are measured. On the plot are some of the famous single-dish telescopes: the 76-meter Jodrell Bank antenna near Manchester, U.K., which gained instant fame when it tracked the rocket carrying *Sputnik I* into space in October 1957; the 64-meter Parkes radio telescope in New South Wales, Australia, which was instrumental in receiving images from the *Apollo 11* mission and relaying Neil Armstrong's "giant leap for mankind" to the rest of the world in 1969; and the gigantic 305-meter Arecibo dish in Puerto Rico, whose spectacular size and location has been featured in such films as the James Bond movie *GoldenEye* and *Contact*, in which Jodie Foster plays an astronomer who decodes signals from aliens.

Of course, radio telescopes have done much more than add scientific spice to pop culture. They see the universe at frequencies that cannot be studied by optical telescopes and thus have been instrumental in the discovery and study of exotic cosmic objects. Quasars (for quasi-stellar radio source) appeared as point sources of light to optical telescopes, indistinguishable from stars; no one really bothered about them until the late 1950s, when radio telescopes picked up strong emissions from them—a discovery that eventually led astronomers to realize that quasars are not ordinary stars. Improved optical telescopes showed them to be extremely energetic galaxies that are beaming out electromagnetic radiation in both the visible and radio frequencies. Radio telescopes also discovered pulsars, which are dense remnants of large supernovae. Invisible to optical telescopes, they showed up as enigmatic emitters of periodic pulses (thus their name) of radio waves.

The push toward greater sensitivity led radio astronomers to come up with new designs. The most important of these used an array of dishes instead of a single antenna. Combining the output of many antennas into one coherent signal is somewhat easier in radio wavelengths than in optical ones, but difficult nonetheless (the British astronomer Martin Ryle received a Nobel Prize in 1974 for developing the technique). Using a number of small dishes allowed astronomers to increase the collecting area without having to face the engineering challenges of building and steering a single massive dish. More collecting area meant better sensitivity, and radio telescopes zoomed up the sensitivity chart. The Very Large Array (VLA) 50 miles west of Socorro, New Mexico, consists of twenty-seven dishes, each 25 meters in diameter. This gives the VLA the effective sensitivity of a single dish about 130 meters across. But more important, the dishes are spread far apart, and this gives the VLA the effective resolution — the ability to see fine details in the sky — of a dish 36 kilometers across. The VLA is almost a hundred thousand times as sensitive as Grote Reber's original radio telescope.

Even more powerful is the Giant Metrewave Radio Telescope (GMRT), an array of thirty dishes, each 45 meters in diameter, about 50 miles north of Pune, India. Steve Rawlings of Oxford University, the former chair of the SKA's international science working group, has called it "the world's leading telescope in the SKA frequency range."

"One of the brilliant things about the GMRT," Rawlings told me, "is that they went for a very cheap dish design and therefore built a bigger telescope than anyone else has, for a remarkably small budget, and there are real lessons for the SKA in that." The GMRT's cheap dish design was the brainchild of Indian radio astronomer Govind Swarup, whom I met with in Manchester, U.K., in 2007, where he had come to accept the third Grote Reber Medal for his contributions to radio astronomy. By the early 1980s, Swarup was getting ready to build a radio telescope in India but did not want to build dishes with solid parabolic surfaces like those used by the VLA: They would be too expensive. "It was very clear that if I followed what the West was doing, I could never compete," he said. In 1986, he hit upon the idea that would help him build a giant—but inexpensive—telescope. The surface of each antenna is made from sixteen tubular steel frames. To these are tied steel ropes, each of which has just the right amount of tension to cause the framework to form a parabolic shape. Swarup then "stapled" a fine wire mesh onto this framework, which serves as the dish's surface. The mesh and the rope trusses are the real innovations, helping the astronomers pull off what has been called "the great Indian rope trick." The entire thirty-dish array cost about $12 million, in the currency of the late 1990s, whereas the VLA cost about $78.5 million in 1972 dollars (the year Congress approved it), equivalent to more than $300 million in the late 1990s. But could the cheaper GMRT, with its crude-looking surface, receive radio signals of the right frequencies?

The surface of a dish dictates the frequencies it can receive. The smoother the dish, the higher the frequency response. A mesh surface was obviously not smooth. What possible good could it do? As it turned out, it was good enough to receive radio waves from frequencies of about 100 megahertz to 1.5 gigahertz—and some of the most important radio signals from the universe lie smack in the middle of this frequency range. These radio waves emanate from the very distant clouds of the most abundant element in the cosmos, hydrogen.

About 370,000 years after the big bang, the universe was transformed. Before this, the cosmos was so hot that matter existed only in the form of electrons and nuclei of hydrogen and helium; the temperature did not allow the electrons and nuclei to form atoms. As the universe expanded and the temperature dropped, the electrons and the nuclei combined, and suddenly the universe became a dense soup of mostly hydrogen and some helium atoms. However, there were tiny variations in the density of this soup—regions where it was slightly thicker than elsewhere. The gravity of these denser regions slowly drew matter together until enough hydrogen gas had accumulated to seed galaxies. These enormous clouds of hydrogen (or their remnants in galaxies) are what modern radio astronomers are after—and the GMRT is uniquely capable of detecting them. Find clouds of hydrogen and you can find galaxies so faint as to be invisible to optical telescopes.

Neutral hydrogen atoms have a telltale signature. The electron and proton that make up the atom each have a spin. The spins can be parallel to each other or antiparallel, and in the parallel state the hydrogen atom has a smidgen more energy than in the antiparallel state. Occasionally the atom will flip from the higher energy state to the lower, emitting a photon with a wavelength of 21 centimeters, or a frequency of 1420 megahertz. This is a radio wave. These radio waves lie outside the range of the GMRT's mesh antennas. But take into account the expanding universe and suddenly these same radio waves are fair game. That's because the 21-centimeter wavelengths emitted by hydrogen atoms in the early universe have been stretched, or redshifted, over time because of the universe's expansion. By the time they reach Earth, they have wavelengths in meters, something the GMRT's mesh antennas are eminently capable of detecting. Today, the GMRT has become a key telescope for studying neutral hydrogen in the universe, which serves as a marker for probing the evolution of galaxies and clusters of galaxies. The GMRT is also proving its worth as a prototype for what the SKA will do with significantly more sensitivity and resolution.

Radio astronomer Subramaniam Ananthakrishnan, who played a major part in building the GMRT, drove me to the site in India, and in an eloquent moment swept his arm out toward the rural landscape dotted with giant antennas and said, "I'm in love with this." He had reason to be. What is currently the site of the world's largest radio telescope for meter wavelengths was once a desolate region. Finding such a site in a land of farmers and substantial poverty had its perils. In 1985, astronomers were surveying a region surrounded by lush sugar cane fields in Dhond, about 60 miles from Pune, when villagers wielding knives and axes surrounded their jeep. "They told us that if we tried to purchase their lands via the government, they would quickly despatch us to heaven," said Ananthakrishnan. "We saw their point and left Dhond in peace." Eventually, the astronomers settled on a patch of land in Khodad, a sleepy village with a less excitable population.

Twelve of the GMRT's thirty antennas are clustered into a square kilometer near the control room. This central cluster gives the telescope its sensitivity. The other eighteen antennas stretch out along three arms, each 14 kilometers long, giving the GMRT its resolution. The telescope opened to international astronomers in the summer of 2002 and, despite considerable glitches, is coming into its own. "The GMRT has been a huge success, really," said Oxford's Rawlings. "We have a fairly big astronomy group here, and we now use the GMRT more than the VLA. Which I think is saying something."

Today you can see cows grazing below the dishes, but this tranquility is in danger of disappearing. What was once barren land has now become the center of India's emerging wine industry, and this brings with it some unusual problems. Farmers occasionally dig up the fiber-optic cables when they are tilling the land. An increasingly prosperous local population means more radio and television stations, mobile phone towers, and power lines, all of which are detrimental to a radio telescope, since radio waves from these sources can swamp the faint signals from the distant universe. This sort of interference is exactly what proponents of the Square Kilometre Array are trying to avoid. The key, as Adrian Tiplady implied, is accessible desolation.

Accessible desolation is the Karoo's unique trait—at least for now. South Africa's chances of winning the bid for the Square Kilometre Array depend in no small measure on preserving this desolation. In October 2007, when I arrived in South Africa, the country's parliament had just approved a bill that would help protect the region for astronomy—not just for radio telescopes but for optical and gamma-ray telescopes. But the farmers who live here have rights, too. They want their mobile phones, televisions, and radios. So the SKA team had to work with phone companies to design special antennas that will provide signals to urban pockets while beaming nothing toward the SKA site. Television towers are another problem. "The Northern Cape is a vast area," Adrian told me. "TV transmitters, in general, will transmit to about a hundred people for every two thousand sheep"—and to reach these few widely dispersed human beings, the transmitters have to be powerful. As an alternative to powerful towers, the newly approved bill would allow the SKA team to work with the broadcasters to deploy low-powered transmitters, in conjunction with satellite downlinks, to provide television to the towns and other remote areas.

Even radio communication from planes flying overhead posed a problem for the SKA. To the north of the proposed SKA site is a flight path, but only a couple of flights travel it each day. However, about 100 kilometers to the south is the extremely busy route between Johannesburg and Cape Town. The new legislation would allow the SKA operators and the aviation industry to cooperate, ensuring that any potential interference from aircraft radars or radios was identified and characterized so that its effect could be eliminated or minimized.

Little is being left to chance in South Africa's bid for the SKA. The competition to host the SKA was announced by the International SKA Steering Committee (ISSC), a consortium of radio astronomers from five European countries and the United States, Australia, South Africa, Canada, India, and China. When the South Africans first announced their intention to compete for the SKA in 2003, a barely concealed titter rippled through the international astronomy community. Surely an African nation stood no chance against the likes of Australia, China, Argentina, and Brazil? It was blatant Afro-pessimism. "We were not taken very seriously to begin with, really," said Adrian.

During our long conversation as we drove through the Karoo, it slowly became clear that the attempt to win the Square Kilometre Array is important to the hopes not just of South Africa but of the continent. So much so that the South African government has thrown its weight behind the bid and other African governments have lent them support. The South Africans want the world — and, indeed, their own people — to understand that their country is capable of high-tech, cutting-edge science and technology. And just so they are not seen as a vast developing country that is merely providing land for the telescope, the South Africans have embarked on an ambitious program to build their own large radio telescope, regardless of whether they get to host the SKA or not. Work is under way on a seven-dish prototype telescope called the Karoo Array Telescope (KAT), to be completed by the end of 2009. By the end of 2012, the KAT will have become the eighty-dish MeerKAT (a play on words, for "meer" in Afrikaans means "more," and the meerkat is an iconic creature of the Kalahari Desert). The MeerKAT will also be a pathfinder, developing, testing, and costing the technology that will be needed for the SKA. The governments and national science-funding agencies that will fund the SKA will announce their choice of the site for the telescope by 2012.

That South Africa is even mentioned alongside Australia as a serious competitor is mostly due to an enthusiastic team led by former radio astronomer Bernie Fanaroff. His journey from scientist to influential, politically savvy administrator is intimately intertwined with South Africa's recent bloody history.

In 1970, Fanaroff—a young white South African—arrived in Cambridge, U.K., to do his Ph.D. Cambridge University had become a heady place for cosmology, thanks mainly to the fight between two resident physicists, Martin Ryle and Fred Hoyle, over the rival theories of the big bang and the steady-state universe. Ryle, a radio astronomer, had discovered galaxies that emitted far more energy in radio wavelengths than in the optical and had embarked on increasingly ambitious surveys to find and catalog these "radio galaxies." He found that there were more of the most powerful radio galaxies per unit volume in the early universe than in the nearby universe, which implied that the universe had changed with time and so was not consistent with the steady-state theory. The proponents of the steady-state universe, led by Hoyle, had predicted that the number of radio galaxies should remain the same regardless of their distance. Ryle proved them wrong with his series of surveys.

Still, the steady-state theorists kept attacking the Cambridge surveys, arguing that the surveys were error-prone and therefore could not be relied upon to prove or disprove theories. So analyzing and understanding radio sources remained deeply important. When Fanaroff came to Cambridge, his advisor, Malcolm Longair, put him to work picking apart the finer details of radio sources. What was the relationship of their numbers to the frequency at which they could be seen? How did the diameter of the sources change with redshift? While peering intently at pictures of these radio sources, taken by the Cambridge 1-mile and 5-kilometer radio telescopes, Fanaroff, along with colleague Julia Riley, realized that the most powerful radio galaxies and quasars had the sharpest lobes at their ends, while the less powerful ones tended to diffuse outward. Based on these insights, the duo developed a classification system for radio sources—called the Fanaroff-Riley classification—that remains in use to this day.

Fanaroff finished his Ph.D. in 1974 and was immediately faced with a decision: to work as an astronomer in Europe or return to South Africa, which was still under apartheid rule. Hedging his bets, Fanaroff applied for jobs in Italy and South Africa. The Italians offered him one, but a postal strike in Italy meant he never received the news. Meanwhile, the University of the Witwatersrand agreed to employ him as an astronomer. Fanaroff returned to South Africa on April 1, 1974.

But Fanaroff soon became frustrated. His real motive for returning to South Africa had been politics, something he had conveniently neglected to tell his university. He had long been inspired by his parents, who had run night schools for black African workers since the 1920s. But Fanaroff found precious little that he could do to bring about reform. He considered returning to Europe. Meanwhile, he started working with a group of academics, students, and workers who were reorganizing unions of black African workers—the previous unions had been dismantled by the apartheid government in the early 1960s. Then, on June 16, 1976, things changed in South Africa. Black schoolchildren in Soweto protested against having to learn Afrikaans. The rally, meant to be peaceful, turned violent when the police fired on the children with live ammunition. The bloody rebellion, which came to be called the Soweto uprising, quickly spread throughout the country. The National Party government reacted in

part by banning all organizers of black trade unions. Fanaroff realized that he could help cover for the organizers who had been black-
listed. He asked his university for two years of unpaid leave, but they
refused, so Fanaroff resigned at the end of 1976 and joined the union
movement full-time.

The apartheid-era government recognized unions for white
workers and workers of mixed race (mainly Indians and so-called
Coloured, the descendants of marriages between white settlers and
Malay slaves or native Khoisans). But unions for black South Africans got no such recognition, leaving them with little or no recourse
when things went wrong. A white foreman could fire a black worker
because he didn't like the man's face, or because he was wearing the
wrong color shirt. A firing could mean dislocation, for black workers
were allowed to stay in "white" urban areas only if they had a job.
"Once you lost your job, you were kicked out, and you had to go back
to some godforsaken hole in the Transkei or Kwazulu, which many
of these guys had never seen in their lives, let alone had a place to go
to," Fanaroff recalled over dinner in Johannesburg. All black South
Africans had to carry passbooks, endorsed by the authorities, which
allowed them to stay in the white areas. A domestic worker, for instance, had such a passbook, and he or she could stay in the backyard
of the white family's home. Often, the worker's spouse and children
would sneak in and sleep together in the backyard. "But the police
used to raid, at three or four o'clock in the morning, and inspect the
backyards," Fanaroff said. "And if they found people there, they'd ar-
rest them and put them in jail."

It was to fight against such injustice that Fanaroff joined the union
movement and was soon meeting black workers in secret, helping
them organize against unjust firings and very low wages. The security police came for him one day in 1977, soon after he started working full-time at a union office in Johannesburg. They took him to the
police headquarters in the city, a building called John Vorster Square,
where he rode a special lift from the garage that went straight up
to the security police offices on the ninth floor. As they were going
up, the policeman escorting him said, "You know, people fall out of

the windows on the ninth floor," referring to the deaths of several detainees. Up on the ninth floor, the police brought out a thick file and said, "We have been watching you." Fanaroff knew they were lying—he had been careful throughout his university career not to get involved in politics. The file had to be on his parents, and this was an attempt to intimidate him. As it turned out, they were trying to confirm that he was getting paid by the anti-apartheid African National Congress underground (which he wasn't), but they had no evidence and had to let him go.

In 1984, the black unions began to work with the United Democratic Front, the aboveground wing of the ANC, and together they started organizing the job actions known as stay-aways. In the first major stay-away, 1.5 million workers stayed home, and the movement grew. "Stay-aways and strikes were a major contributor to making the country ungovernable," Fanaroff said. "Which was a major contributor to the ruling National Party recognizing that they had to negotiate, which eventually led to the release of Nelson Mandela in 1990 and the start of negotiations." Fanaroff fondly remembered the end of apartheid and the first democratic elections in South Africa. "It was the most amazing time. Everybody stood in long queues, and there was virtually no crime for the three days of the election. Everybody was so happy, right across the country. Nobody would have predicted it, because the negotiations were very acrimonious."

When Nelson Mandela formed his government in 1994, Fanaroff was made the deputy director general in the Office of the President and head of the Office for the Reconstruction and Development Programme. Two years later, he was named to run the National Crime Prevention Strategy. After the 1999 elections, a new minister of safety and security wanted to make Fanaroff a divisional commissioner in the police. Fanaroff declined. "I don't take orders well," he told me, "and I would look ridiculous in a police uniform, so I stayed on as the minister's advisor for a year and a half and then left the government service."

Soon life was to come full circle for Fanaroff. In December 2002, he received a phone call from the director general of the Department of Science and Technology, Rob Adam, and the president of the National Research Foundation, Khotso Mokhele. They wanted him to manage South Africa's bid for the Square Kilometre Array. Fanaroff agreed. But they had to move fast, because the other countries had been working on their own bids for years. By May 2003, the team had identified a site in the Karoo and turned in a high-quality proposal, surprising the international radio-astronomy community. But skepticism remained about South Africa's ability to pull off something as ambitious as the SKA. In January 2004, a number of radio astronomers came to South Africa and flew to the site. "There is nothing that dispels prejudice like actually being in a place," said Fanaroff. The bid was so successful that Australia and South Africa were shortlisted in 2006 as the only two places on Earth suitable for the SKA.

Adam, Mokhele, and Fanaroff convinced the South African government to throw its political and financial weight behind the bid. The country decided that it did not want merely to host the SKA but also to build the MeerKAT and turn the Karoo and South Africa into a hotbed of radio astronomy and fundamental physics. (An 11-meter optical telescope is already operating in the Karoo, and the groundbreaking HESS [High Energy Stereoscopic System] gamma-ray telescope is not far away, in southern Namibia.) If South Africa gets to host the SKA, it will further transform science on the continent: A few of the telescope's antennas will be sited in Namibia, Botswana, Mozambique, Madagascar, Mauritius, Kenya, Zambia, and even as far afield as Ghana.

One of the odd things about the Square Kilometre Array is that it might become the last giant radio telescope operating at centimeter and meter wavelengths to be built. Its sheer size means that the best way to build a bigger telescope will be to just add antennas to the SKA. Or build telescopes that look entirely different from the Square Kilometre Array, using technologies that are unimaginable today. The SKA will be magnificent in scale. After years of wrangling, astronomers have settled on using about three thousand small-diameter dishes to receive signals of frequencies ranging from 1 to 10 gigahertz, with the hope of someday extending the upper limit to 35 gigahertz. For lower frequencies, going down from 1 gigahertz to about 100 megahertz, and possibly even down to 70 megahertz, the SKA will most likely use what are called phased arrays.

Phased arrays are not sexy. While radio telescopes like the VLA have seized the popular imagination, with their many parabolic dishes all moving in concert, focusing our attention toward the distant cosmos, the rather mundane and immobile phased arrays have escaped the cult status accorded their dishy cousins. But they have been just as important for radio astronomy as the steerable dish antennas.

One such phased array took shape in the fields outside Cambridge, U.K., in the mid-1960s. Radio astronomer Antony Hewish wanted to look for quasars and other radio galaxies. At the time, quasars were known to be powerful and extremely distant sources of radio waves. Today we know that quasars are young galaxies from the early universe that have supermassive black holes at their centers, but in the 1960s quasars were still a mystery. To look for them methodically, Hewish designed a telescope that covered 4.5 acres of farmland. It had 2,048 dipole antennas, each of which was basically a wooden post in the ground with an aerial mounted horizontally near the top of the post. The telescope couldn't be steered to a spot in the sky like a dish antenna; it received signals only from a tiny strip of sky above Cambridge. The Earth's rotation ensured that the telescope covered a large portion of the sky in a twenty-four-hour period, one small slice at a time.

But within the slice of sky visible to the telescope, the instrument could zero in on specific locations. This is the magic of phased arrays. The signals from each of the 2,048 antennas were combined to get the final signal. The signal from any given antenna could be delayed electronically so that it was ever so slightly out of phase with the signals from other antennas. By adjusting each of the delays, the telescope could blot out some signals and intensify others. Essentially, the telescope could electronically focus on a given location in the slice of sky.

All this depended on wiring the antennas with great care, for the length of wire from each antenna to the control room was critical. The astronomers had to know exactly how long each signal took to reach the electronics; introducing electronic delays would have been futile if each signal's travel time was not precisely established. The job of wiring the antennas fell to twenty-four-year-old Jocelyn Bell, one of six Hewish graduate students helping to build the telescope. When students started in Hewish's group, they were each given a set of tools—pliers, wire cutters, a screwdriver. Bell left the sledge-hammering of the posts to the "blokes" but did the wiring herself. For two years, the students toiled, sometimes in the cold, icy rain of Cambridge with hot-water bottles tucked under their duffle coats and mackintoshes. "It was pretty perishing," Bell said not long ago, as we sat in the Fellows' Room of the Royal Society in London.

The telescope became operational in July 1967. The data were recorded by a pen whose tip moved up and down as a sheet of paper rolled beneath it, much like a seismograph. This was before computers, so Bell had to analyze nearly a hundred feet of data each day, looking for any signals that might reveal a quasar. By the end of October, she had realized that there were some squiggles on the paper that made no sense. They were radio signals unlike any that had ever been seen, but the signals were squished together, as the paper moved only about an inch every five minutes. The only way to know what was going on was to blow up a section of the tracing. Bell did this by running the paper faster under the pen at precisely the time when the strange signals were expected. At that speed, an entire day's worth of paper sped through the plotter in fifteen minutes or so, and the signals were now stretched out. Bell did this for a month but found nothing out of the ordinary. Then, in November, there it was. "It was beyond imagination," she said.

What she had found was a pulsating signal, a short blip that repeated with astonishing regularity. Hewish was a cautious man, so he forced his team to look for the obvious explanations. Was the signal coming from something on Earth? Taxi radios, airplane altimeters, communications satellites? Was something wrong with the telescope, its wiring? Just as with Karl Jansky's discovery, the first clue that they were dealing with something real was that the signal kept sidereal time, appearing four minutes earlier each day. Persistence would once again turn into serendipity.

The team worked out the distance to the source of the radio signal. When radio signals travel through space, the waves interact with electrons in their path. This causes low-frequency waves to slow down relative to the high frequencies. By estimating the number of electrons in space and by measuring the difference in the arrival times of the high-frequency versus low-frequency components of the signal, Hewish and Bell put the source at a distance of 200 light-years. "Which put it beyond the solar system," she said.

Christmas was approaching. Just before heading home for the holidays, Bell found another, similar signal. Unable to do anything more at the time, she went to visit her parents in county Armagh, Northern Ireland. Her father, an architect, took her to see the planetarium he had designed for the Armagh Observatory. Astronomer Patrick Moore, the first director of the planetarium, was showing them around. The building was not yet complete—no seats, just concrete flooring—but the projector was working, and Moore flashed an image of the sky onto the planetarium dome. Bell stood silently, gazing at the dome overhead, trying to find the patch of sky where the first pulsating signal had come from. She had already determined that it was in the direction of the constellation Vulpecula. "I realized what a hell of a lot of stars there were in that bit of sky," she recalled. "But I didn't say anything to anybody, because at that stage we were not at all confident of what we were dealing with."

After Christmas, Bell returned to Cambridge and found two more such signals. Finally it was time to reveal to the broader astronomy community that they had found four pulsating radio signals coming from different parts of the Milky Way. The team had talked among themselves about the signals being from "little green men," aliens on another planet, but in jest. The fact that they had found four such sources and that observations had shown that none of them had come from a planet in stellar orbit—such a signal would have had a characteristic wobble—meant that aliens were not responsible. Moreover, the pulses coming from these sources were very predictable. The first one, for instance, lasted exactly 0.016 second and repeated with even greater precision every 1.33730113 seconds. The others were similar. Something small, about the size of a compact star, was causing them.

Hewish went public with the announcement, suggesting that the sources had to be either white dwarf stars or neutron stars. He called them "pulsating radio sources," which got shortened to pulsars. Within a year, pulsars were confirmed to be neutron stars — compact remnants left behind after massive stars go supernova. Beams of radio waves emanate from the poles of these neutron stars, and when the beams happen to be pointed toward us, they sweep across Earth like lighthouse beacons. These stars, made mainly of neutrons, are unbelievably dense. "The density of that star is like taking a thimble and squashing the whole of the world's population into that thimble," said Bell.

For weeks after the announcement in 1968, Hewish and Bell gave interview after interview, and to her dismay Bell found that these interviews all followed a similar pattern. "They would ask Tony about the astrophysical significance of this discovery, and they would turn to me for the human interest," she remembered. "Which meant asking questions like what were my vital statistics, how many boyfriends did I have at once, was I taller than Princess Margaret. Young women were page-three material. They didn't know what to make of a young woman scientist. And I must say, I didn't know how to deal with the situation either. I forgot my vital statistics, and I was generally uncooperative." The sexism carried over to the photo shoots. "The photographers would look at me and say, 'Could you undo a shirt button or two, please?'"

Bell recounted this with no trace of bitterness, surrounded as she was by paintings in gilded frames of grand old men of science at the Royal Society. Hewish went on to receive the Nobel Prize in 1974 for the discovery of pulsars. Bell, meanwhile, got married and had her first child. ("And it kind of said to me, 'That's the way the world is. Men get prizes and women look after babies.'") But she took pride in the fact that the discovery of pulsars was the first astronomical discovery that had been deemed good enough to get a Nobel Prize in physics. Until then, astronomy had been ignored.

Meanwhile, pulsars continued to astound, even leading to precise confirmations of Einstein's general theory of relativity. Soon after the first pulsar was found, the eminent Dutch astronomer Ed van den Heuvel predicted that one in a hundred pulsars should be a binary—that is, a pulsar orbiting another similarly massive object, maybe another neutron star or even a black hole, in a tight, gravitationally bound dance. In 1974, just about the time that the pulsar count was reaching one hundred, Joseph Taylor, then at the University of Massachusetts at Amherst, and his student Russell Hulse, using the 305-meter radio telescope at Arecibo, stumbled upon a binary pulsar system. The period of the radio pulses from the source suggested that something weird was going on. Hulse was at Arecibo, and he called Taylor, who was in Amherst. The two did some quick calculations and realized that the pulsar was rapidly orbiting another object and that the speeds involved were such that general relativity had to be brought into the picture. Newton's theory of gravity was no longer enough.

Of course, radio astronomers at the time had never dealt with binary pulsars and Arecibo had no books to help them out. So Taylor went to a bookshop in Boston and bought every book on general relativity he could find. The analysis proceeded furiously, sometimes with Hulse dictating his data over a bad phone link from Arecibo, and soon enough they both realized that they had the first way to test the effects of general relativity outside our solar system.

The theory predicted that such a system would generate gravitational waves, or ripples in the fabric of spacetime, because of the intense gravity of these dense, compact objects. The waves would radiate off into space, and the energy lost would cause the two objects to slowly spiral in toward each other. Observers on Earth could measure this effect by studying the time interval between the radio pulses. The pulse period could be used to calculate the time it took the pulsar to orbit its partner, and theory showed that the orbital period should decline by about 75 millionths of a second per year—an astoundingly small number.

Over a span of four years, Taylor and several colleagues measured the changes in the pulsar's orbital period and found it to be in accordance with theoretical predictions. The binary system was indeed losing energy to gravitational waves. For their discovery, Taylor and Hulse won the Nobel Prize in 1993.

Taylor knew Bell very well; both had more in common than just an interest in pulsars. They are both committed Quakers, and the combined interest in science and religion had forged a strong bond. When Taylor went to Stockholm to pick up his prize, he invited Bell to join him as his guest. It was his way of acknowledging Bell's contribution to the field. "I think it made the Nobel Prize committee slightly twitchy," said Bell. "It was making some kind of statement."

These are the experiences that have shaped Bell, and she has spent her career working for a level playing field for women. On a recent trip to South Africa, she spoke about her work to students at a summer school for disadvantaged children in Pretoria. The school was being run by Anita Loots, the MeerKAT project manager. As Bell walked into the school, the kids started singing Nkosi Sikelel' iAfrica, or "Lord, Bless Africa," a song written in 1897 and widely sung during South Africa's struggle against apartheid. It is now South Africa's national anthem. Bell teared up. "[The kids] and their situation seemed to encapsulate for me so much of South Africa today—the deprived background, the ability and potential, the long, hard road ahead."

Bell, ever the astronomer, is looking forward to a new era of discovery with telescopes like the MeerKAT and the SKA. "God knows what they will find, but they will find things," she said. "I'm excited, but I'm also aware that by the time the SKA comes along, I'll be very doddery."

In all likelihood, the Square Kilometre Array will find things that will surprise astronomers. But Steve Rawlings isn't leaving it to chance. He wants to use the SKA for precision cosmology, and the telescope's size and sensitivity will allow him to do just that. It will require more than just waiting for a serendipitous discovery. Rawlings is taking a more modern approach, which means "designing" experiments, as one would for a particle accelerator. He is busy building complex mathematical models of the universe and predicting the distribution of galaxies, with the intent of testing the predictions with the SKA.

Strangely, for a man who is now immersed in astrophysics, Rawlings showed little interest in the stars as a kid. Thanks to a father who had been a navigator in the air force during World War II and knew "all about stars and the rest of it," Rawlings rebelled. "I didn't want anything to do with it," he told me when we met at Oxford in April 2007. But he did enter Cambridge to study physics and ended up hearing Bell's advisor, Antony Hewish, lecturing on cosmology. Influenced by Hewish, Rawlings pursued a Ph.D. in radio astronomy. However, it was all about maths and physics, not stars. "Honestly, I still don't know any constellation beyond the Plough," he said, laughing.

Not that it matters. Rawlings's and the SKA's sights are set way beyond any constellation. Of all the objectives of the SKA (and there are many), the one of most interest to cosmologists is the deciphering of dark energy and measuring the curvature of spacetime. And at the heart of this plan will be the extraordinary capabilities of the SKA. One hundred times more sensitive than existing radio telescopes, the SKA will observe a billion galaxies out to a redshift of 2, when the universe was only about 4 billion years old. The Sloan Digital Sky Survey-III (SDSS-III), when it's completed, will have optical images of about 1.5 million galaxies, using the 2.5-meter telescope at Apache Point Observatory, in New Mexico. The SKA, however, will look for clouds of neutral hydrogen and use them as tracers of galaxies. "The big surveys of hydrogen that the SKA will do will give you the equivalent of the Sloan survey, but with a thousand times more volume and a thousand times more galaxies, and that has huge cosmological implications," said Rawlings.

Until the SKA comes along, SDSS-III will remain the definitive study. In 2004, Daniel Eisenstein of the University of Arizona, Tucson, and the SDSS team detected the imprint of sound waves that rang through the early universe. The only previous evidence of these sound waves had been seen in measurements of the cosmic microwave background. This extraordinary aspect of the universe goes back to the very first instants after the big bang. Inflation had ended. Quantum fluctuations had ensured that certain parts of spacetime had more energy than others. These fluctuations had become regions of high density and low density: The denser parts had relatively more dark matter and normal matter. Then, about 30,000 years after the big bang, the denser dark matter started collapsing inward, and baryons (i.e., protons and neutrons in the form of nuclei) and electrons followed suit. But baryons and electrons, which were tightly coupled to photons, were impeded in their gravitational collapse by the outward pressure exerted by the photons. So as the baryons were sucked in by gravity, they were repelled by the photons. This set off a pressure wave, which behaved in much the same way as a sound wave. A spherical shell of baryonic matter sped outward from each dense spot at about half the speed of light. The matter in the universe started oscillating, giving rise to what cosmologists call baryonic acoustic oscillations. The resulting primordial sound waves were sent ringing through the cosmos.

Fast-forward to about 370,000 years after the big bang, the so-called epoch of recombination. The electrons and baryons combined to form hydrogen and helium. Suddenly, the photons could not collide with free electrons and so were unable to exert any pressure; the oscillations stalled. What remained at this time was an over-dense region of dark matter and a spherical shell of normal matter surrounding it. Of course, there was more than one shell, each centered around its own seed of dense dark matter. The shells overlapped, not unlike the multiple waves that are created when you chuck a number of stones into a placid lake. Theorists realized that these shells of denser normal matter and the dense dark matter at their centers should have left their imprint on the structure of the universe: More galaxies should have formed along the contours of these shells and at their centers. At the time of recombination, the radius of the biggest shells had a fixed size: the primordial ruler. Taking the expansion of the universe into account, the ruler today would be 450 million light-years across. If one were to look for galaxies in a large volume of space in the nearby universe and examine their distribution, the shells and their centers should become apparent. This is what astronomers did. They took one galaxy and then probed the distance at which another galaxy was most likely to be found. They had no idea whether they would find pairs of galaxies that are 100 million, or 500 million, or 700 million light-years apart. As it turned out, the SDSS survey of about fifty thousand galaxies found a slight excess of galaxy pairs that are 450 million light-years apart, just as theory predicted. They had seen the ripples created by the striking of primordial gongs.

Rawlings is looking to more than upstage the Sloan survey. With the SKA, radio astronomers can look further back in time and also study a much greater volume of space. A billion galaxies will illuminate the primordial ruler, and the ruler can be used to better understand dark energy. Theory tells us exactly how big the primordial ruler should be at each epoch of the universe. For instance, in the cosmic microwave background, which was emitted at exactly the time when the oscillations stalled, the biggest ripples have an angular separation of about 1° in the sky. That's the size of the ruler when the universe was 370,000 years old. The SDSS survey of the nearby universe gives us another snapshot: The ruler is 450 million light-years across today.

By probing a billion galaxies all the way back to when the universe was between 3 billion and 4 billion years old, Rawlings hopes that the SKA will give cosmologists the best possible data set to see how the ruler has changed over time and therefore help us to understand the expansion history of the universe. Physicists work with mathematical models of the universe based on the general theory of relativity. Parameters in the models include the curvature of spacetime (is it flat or curved?), the dark-matter density, the dark-energy density, and the equation of state for dark energy: w, which describes the change in dark-energy density over time. The models are such that tweaking the value of one parameter often affects the value of the others, so cosmologists do their best to fit the available data to their models and constrain the values of these parameters. For instance, they can use the models to calculate the size of the primordial ruler at different redshifts and use observations to verify the model and its predictions. And that's where the Square Kilometre Array will excel. The data it will gather will give us some of the tightest limits on the parameters. Will the SKA's findings confirm a flat spacetime? And given a flat (or curved) spacetime, will the measurements agree with models in which w is equal to -1, meaning that dark-energy density has remained constant? Or will they fit models in which dark-energy density varies over time (making w greater than or less than -1).

The theory that stands to gain or lose the most from precise measurements of the curvature of spacetime is string theory—the best bet for a theory that combines general relativity with quantum mechanics. String theory began to emerge in the late 1960s, when experiments involving hadrons (such as protons and neutrons) were clearly showing that these particles were composite, not elementary. In other words, they were made of more fundamental matter. But physicists were struggling to explain the scattering of hadrons off one another in particle colliders. Then, in 1968, a young Italian physicist named Gabriele Veneziano wrote down an equation that described the scattering perfectly. Still no one, including Veneziano, realized the physical significance of this equation until 1970, when Leonard Susskind, Yoichiro Nambu, and Holger Nielsen, independently and almost simultaneously, figured out that the equation was describing the dynamics of an elastic string. If the elementary particles inside hadrons could be modeled as vibrating strings, then their interactions could be explained as the scattering of one string off another. That was the beginning of string theory, even though it was a rather rudimentary picture of the constituents of hadrons tied together by rubber-band-like strings.

String theorists began showing that different modes of vibration of these one-dimensional strings could give rise to the entire family of particles—both the force-carrying bosons and the fermions that make up matter. Despite the promise of being able to describe the underpinnings of matter, theoreticians kept bumping up against an uncomfortable fact. The hadron-like particles described by these strings kept behaving as if some of the forces acting on them were long-range gravitational forces, whereas the forces that act on the constituents of hadrons are extremely short-range. In 1974, physicists John Schwarz and Joel Scherk realized that some of the vibrating strings were indeed describing a particle responsible for the gravitational force—the hypothetical graviton. String theory, it seemed, had the makings of a theory that could describe all the fundamental forces of nature, including gravity.

One fascinating outcome of the theory is that the strings have to vibrate in ten dimensions: nine of space and one of time. This of course does not deter physicists. Extra dimensions have existed in physics since Einstein's day. In the case of string theory, six of the nine dimensions of space are thought to be so small as to be undetectable. Scientists thought string theory's elegant equations would sooner or later be solved and from them would emerge a description of our universe—a natural explanation for the particles and the forces we see around us, along with a rationale for the values of fundamental constants (such as the mass of the electron). The hope was that string theory would predict the existence of a unique universe: ours.

That hope has been shattered over the past decade. It has become clear that the extra dimensions of string theory can be hidden in as many as 10^{500} ways, if not more. And each way of compacting the extra dimensions leads to a different universe, with unique values for the fundamental constants, and even unique laws of physics. Nothing in the theory—at least at this time—suggests that any one of these ways is preferred over the others. All of them, in principle, are equally likely. In other words, all of the 10^{500} universes are possible. Ours is just one of this mind-boggling multitude. In 2003, Susskind, who wholeheartedly accepts this finding, coined the term "the landscape" to describe the situation.

The physical realization of the theoretical landscape gives us a multiverse. But how can one believe in the multiverse? How will we know if it exists? The first step is to identify certain predictions of string theory that can be tested. In the context of cosmology, some models of string theory predict that spacetime will be ever so slightly curved. As noted, current measurements of the curvature show it to be zero, within limits of experimental error—which means that spacetime is flat—but there is enough wiggle room in the measurements to accommodate a slightly negative or positive curvature. The Square Kilometre Array, when it is working at its full potential, will be able to measure this property of the universe with unprecedented precision. As we will see in chapter 10, the models of string theory that suggest a multiverse also predict that the curvature of spacetime should be slightly negative—so such a measurement by the SKA will be a big boost for these ideas, whereas if the SKA finds a positive curvature it would cast doubt on string theory's prediction of a landscape.

Given the enormous stakes, Rawlings isn't merely interested in pointing a telescope at the skies and waiting for an unexpected scoop. His team is thinking up experiments for the Square Kilometre Array which will help to clarify the nature of dark energy and measure the curvature of spacetime. But he notes that within the SKA community there are traditionalists, with a more romantic view of what a telescope can do. "There are people like me who think of the SKA as an experiment that we have to do which might never work, for whatever reason," he said. "And on the other end are the people who say, 'We can't possibly imagine what the SKA is going to do or see. This thing is going to be a hundred times more sensitive than anything we have. It's going to find something far more interesting than this dark-energy rubbish.' And they are right. The history of radio astronomy is like that."

Out in the Karoo, Adrian finally got his wish for home-cooked Karoo lamb in a B&B in the Northern Cape town of Carnarvon, where we stayed for a night. We left Carnarvon early the next morning, soon turning off the paved road onto a by-now-familiar gravel road. We were nearly eight hours from Cape Town. The site for the Meer-KAT and the SKA was still about two hours deeper into the Karoo. As we drove, the shrubs got smaller and the landscape drier. Occasional splotches of bright pink and white flowers shocked with their incongruity. At times, dense fields of blackened boulders dominated the landscape on either side of the road. They looked as if they had been charred in a fire, but they were in fact chunks of dolerite, weathered to a dark sheen by wind and sun. We also passed massive piles of rocks that seemed to have been laboriously stacked on top of one another. Sometime during the Jurassic period, when dinosaurs roamed the land, vast outpourings of basalt covered the countryside. The basalt eroded, exposing dolerite, which is basalt that cooled and solidified well below the Earth's surface. As the dolerite cooled, cracks formed in the rocks, mostly at right angles. Eventually rainwater flowed along these cracks, eating away at them. The once sharp edges gave way to rounded contours. Today, they look like sacks of wool, hence the term "woolsack weathering" for this geological process.

Adrian braked to show me another odd sight—a nest of weaverbirds on an old telephone pole. In a land devoid of trees, the weaverbirds colonize any freestanding structure. In the Karoo, these happen to be abandoned telephone and electricity poles, and now and then the stump of a former windmill. The nests looked like lumps of dry grass that someone had tossed on top of these poles, but they were carefully constructed, designed to protect the eggs and the young from the weather and birds of prey—at least one of which was circling the nest as we watched.

The abandoned utility poles got us talking about the power requirements for the Square Kilometre Array. The SKA will need hundreds of megawatts of electricity, but bringing this power to the core site is going to be tricky. The best way to transmit huge amounts of power is to use high-voltage lines, but these are undesirable around a radio telescope. High voltage can ionize the surrounding air, causing what's called a corona discharge, which generates radio frequency interference (RFI) strong enough to mess up the faint signals from the cosmos. Even low-voltage power lines can spark, another source of RFI. The solution is to bring power on high-voltage lines to within about 45 kilometers of the core site and then step down the voltage and transmit the power using multiple cables strung on steel towers, finally burying them underground within 6 kilometers of the site. Weaverbirds will have to stick to their old broken poles for nesting.

We turned off onto yet another gravel road. Within minutes, I could sense that we were heading into special territory. In front of us, a range of mesas loomed large, their flat tops made of slabs of dolerite, which had prevented the softer rock underneath from erosion. These mesas will protect the SKA site from the radio transmissions of towns to the south—allowing the astronomers the relative luxury of being near civilization yet protected from it. Accessible desolation. Farther north, the Karoo becomes flat and even emptier, and the SKA can be expanded in this direction if necessary.

Adrian stopped at a metal gate. We were at a farm. A small dog barked shrilly and incessantly at us, but kept backing away as we walked into the farmer's garden. Jan Louw, the farmer, came out to meet us, and as soon as we shook hands, the dog stopped barking. One of Jan's farms was going to be part of the site for MeerKAT and, if all went well, the SKA. He gave us directions to the farm, and he and his dog followed us in a small pickup truck.

There were no gravel roads, only a rutted path created by Jan's truck. Dry, thorny bushes scraped the sides of our dusty 4 x 4. "This is beautiful countryside," said a besotted Adrian. About two hundred years ago, the San Bushmen of the Kalahari occasionally came this far south. Back then, herds of springboks, numbering in the millions, took days to cross a patch of land, raising dust that swirled for weeks. Eventually the Boer settlers arrived, fenced off their immense farms, and hunted down the springboks. The San retreated into the Kalahari. Today, only a handful of springboks grazed around us.

Jan took us to a flagpole stuck in the middle of the site, which is some 90 kilometers west of Carnarvon. This will be the center of the eighty-dish MeerKAT. If South Africa wins the bid for the SKA, its three thousand antennas will be centered here too. The telescope's core will span a circle 5 kilometers in diameter, subsuming the Meer-KAT in the process. The flag—really just a grain sack tied to the pole—once fluttered but had since disintegrated in the relentless sun. Jan tore off a piece of the remaining fabric and crushed it into powder with his fingers—such was the effect of the UV rays in this beautiful but hostile land.

We drove over to a *koppie*, a small hill, not far from the flagpole. The hill was essentially a large mound of dolerite rocks, with bits and pieces of shale strewn around. Standing on this ancient land, the wind whistling in my ears, I saw an endless expanse of shrubs and rocks stretching out toward the north, dry and glistening at the same time. Jan's farm was the only sign of any human activity. He had several thousand sheep, but none were in sight. A few kilometers from where we stood was another hill, dotted with trees—or what passed for trees in this land of shrubs. They were *Kokerbooms*, a species of aloe. San Bushmen used to gouge out the soft insides of the trunk or large branches, tie a piece of leather around the hollowed-out bark, and use it as a quiver for their arrows. It was the only one of two hills in the region that harbor these giant succulents. Some were a few meters tall, and given how slowly *Kokerbooms* grow they had to be several hundred years old, possibly brought here by the Bushmen themselves.

Dry *Driedoring* (three-thorn) shrubs covered the entire landscape. Jan said that the region was still in the grip of a prolonged drought. When it rains here, in short bursts of unpredictable violence, water flows through the flat land before us and the three-thorn shrubs flower, carpeting the land in white. The *Kokerbooms* turn green almost overnight. I stood for a moment, trying to imagine a colorful valley dominated by the rainwashed antennas of the Square Kilometre Array, their gleaming surfaces scanning some far off region of the universe, billions and billions of light-years away.

The Karoo promises to be a significant part of cosmology's evolving story. But there is another land, far bigger than the Karoo, that is already writing its own narrative. About 50 million years after Africa split off and drifted northward from Gondwana, another bit drifted south and ended up at the bottom of the world. A land more pristine than the Karoo, Antarctica is fast becoming our most sought-after window on the cosmos, though it remains as dangerous for the untrained traveler as when humans first started exploring it a century ago. It was with great anticipation that I set off for the frozen continent—cosmology's biggest and boldest bet yet.

7 · ANTIMATTER
OVER ANTARCTICA

BEFORE I SET OUT for Antarctica, I kept hearing expressions like "See you on the ice" and "So how long will you be on the ice?" "The ice," I realized, was an insider's term, a tacit acknowledgment of our inability to describe Antarctica in anything but the most mundane language. Francis Spufford claims, in the introduction to his anthology of writings about the frozen frontier: "Nowhere else on Earth is it so clear that a place has an integrity apart from what we might say about it. Nowhere are words so obviously ineffectual a response to what just, massively, exists, whole and complete and in no real need of translation."

It was only on the continent that I got my first visceral sense of "the ice"—of its unsullied beauty and unforgiving danger. I was taking part in a snow-survival course conducted by the U.S. Antarctic Program at McMurdo Station, mandatory for scientists and staff in the field. Sixteen of us were out on an ice shelf. After a tiring day of learning how to build an igloo-like "Quincy hut," make "deadman" snow anchors to keep tents from blowing away, and saw blocks of dense snow to use in building wind barriers, most of us wisely chose to sleep in the sturdy Scott tents or smaller mountain tents made available to us. I opted to sleep in a snow trench, and in a moment of misguided adventurism I decided to dig the trench myself. Shovel in hand, I gouged out a coffinlike cavity about three feet deep, narrow at the top and wider at the bottom. It was just roomy enough to accommodate an insulating mat and sleeping bag. I crawled into the bag around midnight. Clouds had dimmed Antarctica's twenty-four-hour summer sun. I willed myself to sleep, despite a throbbing headache brought on by the digging.

An hour later, I found myself staring at a narrow patch of glaringly bright sky. The walls of the icy trench seemed to be closing in, and a claustrophobic panic enveloped me. I stripped off the nooselike gaiter from around my neck, shot out of the trench, and stood on the ice with feet clad only in socks, which immediately got cold and wet. I goose-stepped on the ice until I managed to put on a dry pair of socks and insulated rubberized boots, and then looked around. The others were inside their various tents and huts. I couldn't put up a tent now, or crawl into someone else's—it just didn't seem appropriate in a survival course—and I didn't want to return to the trench. All I could do was walk about until the panic receded. The only sound was the crunch of snow under my feet.

I was on the Ross Ice Shelf, named for its discoverer, Capt. James Clark Ross of the Royal Navy, on an 1841 expedition. The region is known for its violent winds, yet to the north—in the shadow of two giant volcanoes, Mount Erebus and Mount Terror (named after Ross's ships)—is a stretch of ice called the Windless Bight. Erebus, the larger of the two and Earth's southernmost active volcano, rises some 12,500 feet from the shelf, with a presence more dominating than most mountains twice as high. The clouds had disappeared, and the sky was a startling blue, against which Erebus's snow- and ice-covered flanks stood out in muscular relief.

The Ross Ice Shelf is the world's largest—nearly the size of France. From where I stood, it looked endless, a staggering sea of white, its icy ripples glinting in the sun. Though my fellow campers were sleeping all around me and our instructors were in a heated hut nearby, I felt entirely alone and not a little panicked. The knowledge that I was in no real danger didn't help. Eventually the stillness of the ice, its quiet immensity, calmed me. I got back into the trench. Inside, I thought of Antarctic explorers who might have been saved by such a trench, which afforded protection against the harsh winds. Maybe claustrophobia was a luxury. Would I have felt this way if my main concern was staying warm and alive? With chastising thoughts of infinitely hardier explorers, I fell into a fitful sleep.

Our survival camp was being conducted during the austral summer, a relatively benign season. The accounts of early explorers reveal just how treacherous it can get here in the winter. Three members of Robert Falcon Scott's 1910–1913 *Terra Nova* expedition journeyed in midwinter through this very part of the ice shelf. In 1911, Apsley Cherry-Garrard, Edward Wilson, and Henry Bowers trudged toward the Windless Bight in "subzero stillness," on their way to a rookery of Emperor Penguins, which nest only during the depth of winter. Their aim was to collect penguin eggs for scientific study. The three walked in almost complete darkness, with only the moon for occasional company. They dodged fields of crevasses and hauled sledges through knee-deep snow. The temperatures were so low that their sleeping bags, wet from the moisture of their own sweat and breath, would freeze solid; it would take an hour of tough toil just to make them pliant enough to sleep in. Even their clothes occasionally froze around them. "If we had been dressed in lead we should have been able to move our arms and necks and heads more easily than we could now," Cherry-Garrard wrote in his 1922 memoir, aptly titled *The Worst Journey in the World*.

They crossed the Windless Bight and reached the slopes of Mount Terror, where they camped. The edge of the Ross Ice Shelf — then known as the Barrier — stretched out ahead of them, its precipitous cliffs plunging down to the sea. They saw gigantic pressure ridges, zones of crumpled ice that form when sea ice pushes against the ice shelf or the shore. "Beyond was the frozen Ross Sea, lying flat, white and peaceful as though such things as blizzards were unknown . . .

Behind us Mount Terror on which we stood, and over all the grey limitless Barrier seemed to cast a spell of cold immensity, vague, ponderous, a breeding-place of wind and drift and darkness. God! What a place!" The trio found the rookery and made it back to camp after more than five weeks, with a handful of eggs, having endured the longest journey yet undertaken in the Antarctic winter.

Scott had set up camp on a point just north of the ice shelf, on the west side of hilly, volcanic Ross Island; today this corner of the island is the site of McMurdo Station, a bustling American base of about a thousand people. Ross Island, some 800 miles from the South Pole, is bounded on the west by McMurdo Sound. The southern edge of the sound butts up against the Ross Ice Shelf, and to the north it opens into the Ross Sea. The Royal Society Range, a small segment of the Transantarctic Mountains, rises steeply from the western coast of the sound, its jagged peaks juxtaposed against the more rounded volcanic hills and islands nearby.

Ross Island gained prominence when Scott's *Discovery* team arrived in 1902 for its first expedition. The team built a hut there which stands to this day. Scott failed to reach the South Pole on his maiden attempt. Ernest Shackleton, who was a member of the expedition, returned to Antarctica in 1908 and built his own hut at Cape Royds, nearly 20 miles to the north, as it was considered unsporting to use Scott's hut. Shackleton, now leader of his own team, came agonizingly close to reaching the South Pole—within 97 miles—but had to turn back. The pole remained unconquered.

Scott returned to McMurdo in January 1911, traveling south again a year later, in the austral summer of 1911–12, with a five-man team. And though they reached the pole on January 17, 1912, a Norwegian team led by Roald Amundsen had beat them by a month. Nearing the pole, Scott found a flag and the remains of Amundsen's camp. Gravely disappointed, he wrote in his diary, "Now for the run home and a desperate struggle. I wonder if we can do it." As it turned out, they couldn't. Two of Scott's men died on the return trek, and Scott and his two remaining companions froze to death in their tent in a blizzard, unable to cover the final 11 miles to their One Ton depot, about 150 miles from McMurdo, where they had left plenty of food and fuel. Scott, a prolific notetaker, wrote his last words on March 29, 1912: "We shall stick it out to the end, but we are getting weaker, of course, and the end cannot be far. It seems a pity, but I do not think I can write more."

Scott was criticized for many things, including what many thought was a misguided pursuit of science when lives were in danger. When a search party reached Scott's tent the following November, they found more than 30 pounds of geological samples, which must surely have slowed the team down. People were left wondering whether Scott and his men would have made it to their depot if it hadn't been for the rocks. Nevertheless, the fossils found in those rocks gave scientists the first glimpse into Antarctica's past, setting in motion the spirit of scientific inquiry that pervades the continent today. As Spufford notes: "Carrying [the rocks] along was, perversely, among the most forward-looking things Scott ever did. It anticipated the coming time when scientists, not explorers, would be Antarctica's defining inhabitants; when understanding, not surviving, would be the most pressing human business there."

While Scott might have anticipated the decoding of Antarctica's past, he could not have imagined the continent's role in deciphering our universe. But he would have appreciated the rationale. On average, Antarctica is the coldest, windiest, highest, and driest continent on the globe, exactly the conditions that have made it crucial to cosmology. The ice sheets that blanket Antarctica are being used as gigantic detectors for neutrinos. The South Pole, with its dry atmosphere and high altitude, is ideal for telescopes. But perhaps most romantic are the high-altitude experiments using giant helium balloons launched by NASA from the Ross Ice Shelf. These flights are designed to study the cosmic microwave background, cosmic rays, and even elusive antimatter particles, and they are made possible by Antarctica's unusual atmospheric conditions.

Every austral summer, the winds in the upper stratosphere over Antarctica settle into a circumpolar anticyclone, which means that at

altitudes of about 36 kilometers they start moving in a counterclock-
wise direction around the South Pole. If you can get a balloon that
high, it will circle Antarctica, and once the experiment has been completed the payload can be brought down over land, not sea—something that is possible (for polar flights) only here in Antarctica: Similar winds form over the Arctic Circle, but they are far less stable,
because they pass over water as well as land. Moreover, since most
of the Arctic Circle is filled with broken sea ice in the summer, a descending payload could easily end up in the drink. Also, the Russians
have often refused permission for flights over their territory, which
is unavoidable if balloons have to circle the North Pole.

Stratospheric balloons can, of course, be launched in the interiors
of continents, but then they are subject to cycles of day and night,
which cause them to expand and contract and thus rise and drop
sharply in altitude, which compromises the experiments. Nor are the
winds above other continents circular; balloons can drift far from the
launch site, making their recovery difficult. And with a few million
dollars being spent on each experiment, retrieval is essential.

Not that retrieving payloads in Antarctica is easy. Though they
come down on land, it's no ordinary land. You need special skills to
tackle the challenges of the ice-covered terrain. Phil Austin has the
right stuff. During my 2007–2008 visit, he was the U.S. National
Science Foundation's camp manager for NASA's Long Duration Balloon (LDB) facility at McMurdo. A hardy fifty-eight-year-old Australian American, he served in the Australian Air Force, then spent
many years "bumming around" the United States, conducting ski
and mountain-climbing tours. In 1982, he arrived in Antarctica for
the first time, to help run snow-survival courses. Years later, he got
a job managing a winter camp for scientists in the middle of Green-
land. That stint led him back to Antarctica, to the high-altitude Russian camp in Vostok, where he helped U.S. scientists conduct air-
borne geophysical surveys of the subglacial Lake Vostok, which lies
4 kilometers under the ice. Vostok lays claim to the coldest recorded
temperature on Earth: –128.6°F. (Compared to Vostok, McMurdo
in summer is relatively balmy, averaging around 20°F.) Austin was

persuaded to take a job providing logistical support for the balloon launches.

During my stay at McMurdo, Austin often talked about what happened once the balloons and their scientific payloads had been launched. Usually, each balloon goes around Antarctica once, or at most twice. Depending on the winds, the flight can last up to three weeks. The flight is terminated by detonating a radio-controlled explosive that severs the balloon from its payload. While the balloon ruptures and falls to ground, the payload, which can weigh more than 2 tons, is brought safely down to Earth by a giant parachute, which opens after a free fall of tens of thousands of feet and separates from the payload just before the landing. Where the payload lands and what happens to it is partly a matter of luck, no matter how precisely the scientists and ground crew have planned the mission.

Austin recounted a couple of particularly difficult recoveries. One was the Boomerang experiment, which flew in 2003 to measure fine features of the cosmic microwave background. "It was a difficult recovery," said Austin. "It ended up [on the Antarctic Plateau] at twelve thousand four hundred feet." The McMurdo team had to use an LC-130 Hercules plane to drop drums of fuel along the route to where Boomerang had landed. Then a smaller Twin Otter, which could land at the site, flew in, refueling on the way, and retrieved the payload over multiple trips.

But the Boomerang recovery was nothing compared to the one that Austin had to coordinate in 2006. An instrument called BLAST (for Balloon-borne Large Aperture Submillimeter Telescope) had come down near McMurdo. This time the parachute had not separated from the payload, and after the landing, winds dragged them both for 125 miles, creating a long furrow in the snow, until the whole thing fell into a crevasse. Austin's team flew over the wreckage in a Twin Otter and took pictures, which failed to reveal the module that carried the hard drives containing the collected data—it was nowhere in sight. The team went back the next day in a Basler BT-67, a converted DC-3, which could fly a little slower and a little lower. The Basler crisscrossed the crevasse and photographs were taken from every conceivable angle. Back in McMurdo, the pictures again disappointed—still no sign of the data module. "At that point, they were pretty much saying, 'OK, we have no payload, we have no data, we have no science,'" Austin said.

The Basler is a big plane, and it had to take wide, sweeping turns to keep coming back over the crevasse. The co-pilot remembered that he had seen something glint on one of those wide turns, so the team went back again and took yet another set of pictures. There was the data module, lying in the snow. Painted white (to help disseminate heat at flight altitude), it had mercifully detached before the parachute had dragged the payload into the crevasse. "Then it was, 'How do we go and get it?'" Austin recalled. "Because it was in this big crevasse field." The plan called for two people to ski, roped together, over 30 miles of crevasses. Austin recused himself and arranged for two younger field-safety training instructors to be flown to the edge of the crevasse field for the long, hard ski to the crash site. Fortunately the pilot found a landing spot within a few miles of where the box had fallen. "The two skied in and found the data module, threw it in a backpack, brought it back here, and said, 'Here you go, folks. Just saved your bacon,'" said Austin.

The recovery of NASA's TIGER (Trans-Iron Galactic Element Recorder), a cosmic-ray experiment on its first Antarctic launch in 2001, was still more harrowing. TIGER did not disconnect from its parachute either, but instead of being stuck in a crevasse, the payload continued to cruise at around 10 knots over the ice, dragged by the fully inflated parachute. The plane with the recovery crew would taxi out in front of the parachute and the men would get out and run toward the billowing chute, hoping to cut the cables. "But it wasn't going straight, it was doing this," said Austin, waggling his arms in imitation of a skier slaloming down a slope. "And you are running toward it, and all of a sudden it's coming toward *you*, and you are running for your life, away from the goddamn thing, on the ice shelf!" Finally the wind died down enough to collapse the parachute, and the payload dug into the snow and stopped—just 15 miles from where the ice shelf met the ocean. Whereas the crew managed to collect that payload, TIGER seems jinxed: The payload from its subsequent flight has never been fully recovered.

I was at McMurdo to witness the second flight of the Balloon-borne Experiment with a Superconducting Spectrometer (BESS). BESS is attempting to answer one of the most perplexing mysteries in cosmology: Why is our universe made of matter and not antimatter? After all, theory says that both matter and antimatter should have been created in equal amounts in the big bang. But matter abounds—in the form of stars, galaxies, planets, and people—while no one has ever detected a single particle of primordial antimatter. BESS is looking for one.

My trip had been timed to coincide with a small window in the austral summer when atmospheric conditions in Antarctica are optimal for balloon launches. The flight from London to Christchurch, New Zealand—the departure point for all U.S. flights to McMurdo—had taken more than twenty-four hours. I was traveling under the aegis of the U.S. Antarctic Program, along with other scientists and staff. The day before our flight, we had to collect our extreme-cold-weather (ECW) gear, which included a bulky red hooded parka, heavy rubberized and insulated "bunny boots," several pairs of gloves and socks, and other articles of warm clothing, such as wind pants, thermal underwear, balaclavas, a neck gaiter, and fleece wear. We had to wear or carry on board gear weighing 20 pounds—all told, the gear weighed nearly 30 pounds—and a few of us wondered aloud if this was in case of a possible emergency landing. But given that most of the journey was over the Southern Ocean and the bunny boots alone would ensure a watery grave, that was unlikely.

Arrival, however, was not guaranteed. Weather conditions at McMurdo can change in a flash, and planes have been known to take off and "boomerang" back to Christchurch. The record, we were told, was seven boomerangs. The hotel staff said they would come knocking early in the morning if there was word of a canceled flight. No one knocked, so it was a go for our 9:00 A.M. departure.

We turned up at the Antarctic Passenger Terminal at Christchurch airport. The weather held, and soon we were on our way in a giant C-17, a U.S. Air Force cargo plane. Its cavernous cabin had comfortable seats for about forty people. Every wire and fuel line was in plain view, and everywhere you looked there were devices to strap things down. There were just two tiny portholes in the front of the plane. It was through one of these that I got my first look at the Southern Ocean near Antarctica, a jigsaw puzzle of bright-white ice floes in deep-blue waters. The plane flew across a stretch of the Transantarctic Mountains, which divide the east from the west side of the continent. Glacier tongues lapped at the sea. Soon we landed on a runway of compacted snow and ice at the Pegasus airfield, which is on the Ross Ice Shelf, south of McMurdo Station. I took my first unsure steps on the shelf with snow blowing across my face. In the haze, I glimpsed mountains around us. We were herded onto a shuttle, a boxy red vehicle on gigantic balloon tires named Ivan the Terra Bus, and it crunched its way across the shelf and over sea ice to McMurdo.

Despite its exotic location, it has often been said that McMurdo Station resembles a mining town in Montana or Wyoming. Bulldozers, packing crates, shipping containers, and prefab buildings all help create that impression—not to mention a hilly terrain of volcanic gravel. About a thousand people are based here in the summer; the population dwindles to a skeleton crew of some two hundred during the winter. At the heart of the station is Building 155, a two-storey, warehouse-like structure where everyone gathers for breakfast, lunch, and dinner. Within a couple of days, it became clear that McMurdo's pulse could be felt in the dining area. It was here at dinner, a week before Christmas, that I met Austin and heard about the balloon launches.

Things were looking good for the first upcoming balloon flight—which would be launching a University of Maryland experiment called CREAM (Cosmic Ray Energetics And Mass) into the stratosphere. "We're mobilizing tonight," said Austin, between mouthfuls of dinner. "We'll meet at DJ at nine o'clock."

DJ was Derelict Junction, a small wooden shed at the shuttle stop in the center of McMurdo Station. A row of corrugated-steel dormitories, painted a drab brown, were on one side of DJ. If you looked down toward McMurdo Sound from DJ, your eyes first rested on a white wooden chapel with a steeple. Farther along was the sound; on its far coast loomed the volcanic Mount Discovery and the Royal Society Range. I was half an hour early, so I wandered toward the coffee house, which used to be the officers' club when McMurdo was run by the U.S. Navy. It was a rusty Quonset hut, also made of corrugated steel, but inside was a dark, cozy bar that could have been anywhere, except for the old skis, snowshoes, and sleds that decorated the walls. I sat inside over an oversize cup of coffee, and waited for Ivan the Terra Bus. It showed up on time and we all got on, wearing our regulation red parkas, for a night out on the ice. "Hold on tight," the driver warned. "The transition is real bumpy." She was referring to the passage from the island onto the ice shelf. A week of relatively warm weather had caused the transition to rot, creating huge pools of meltwater, which grew bigger every time the bus's balloon tires plunged in and out of them. These were Antarctica's potholes. There was no real cause for concern, as the ice shelf is about 50 feet thick near the coast, but the ride was rough, occasionally lifting the unbuckled among us up off our seats.

We arrived at the Long Duration Balloon facility an hour or so later, around 10:00 P.M. Close by is Williams Field, the U.S. Antarctic Program's principal airfield, named after Richard T. Williams, a young driver with the U.S. Navy who died in 1956 when his 35-ton Caterpillar tractor plunged through the ice. He had been part of Operation Deep Freeze, an ambitious navy operation that set up the permanent base at McMurdo and stations at the South Pole and elsewhere in Antarctica. Williams had just crossed a timber bridge over a large crack when the ice gave way. In her book *Deep Freeze*, Dian Olson Belanger recounts what happened:

> The tractor was gone in a moment. Its door and hatch were open. Someone saw Williams partway out, but either an obstacle or the cold arrested him. Standing there, horrified, was

builder Charlie Bevilacqua, who remembered stripping off his heavy clothing and scrambling out onto the broken ice to pitch chunks aside, heedlessly struggling to enlarge the hole so Williams could resurface.

But Williams never did.

The balloon facility is a five-minute walk from Williams Field, on icy snow paths crisscrossed with Caterpillar tracks. An assortment of buildings on skis make up the launch facility. The galley caught my eye: a Korean War–era Jamesway hut, a semi-cylindrical structure resembling a Quonset hut but covered in insulated canvas. These buildings were dominated by two massive hangarlike structures housing the payloads — the science experiments. A big red truck with a towering black crane was parked outside one of the buildings, holding a payload that was destined for the stratosphere. A few minutes' walk from the buildings was the launch pad, a circular patch of compacted snow about 1,600 feet across. We were farther out on the ice shelf than we had been at the snow-survival campsite. The ice here is more than 150 feet thick. Mount Erebus seemed even more majestic at this distance; today, white plumes of steam hovered above its summit. Liftoff for CREAM was still hours away, and the slow-moving plumes above Erebus added to the feeling of suspended animation as we awaited the early morning launch.

The launch depends on the surface winds, which have to settle down to less than 10 knots, even at 5,000 feet, and blow in the same direction; otherwise, getting the gigantic 1,000-foot-tall balloon off the ground without seriously endangering the payload and the ground crew is impossible. The winds, however, were still whipping around at speeds of 15 to 20 knots. I walked back to the galley to wait it out. There were coffee and pastries, and the cook, an amiable young man named Matt, had made a delicious pumpkin bisque. Hot soup and croutons on a freezing ice shelf — it was an entirely pleasant way to bide one's time.

Well past midnight, I walked out to see how the scientists were doing with the payload, which was now hanging off the launch truck.

There was no one there except for one of the crew. (The balloon is of course unmanned, but a small, highly trained team is required to get it off the ground.) Normally, he'd be seated, along with another crew member, holding on to a rope tied to the payload to prevent it from swinging in the wind. This time he was alone and on his feet, pulling at the rope. He saw me and beckoned frantically. "Tell them the wind's too strong," he shouted. "I can't hold it."

I ran toward the rigging building and let out a shout of my own, and soon a few people turned up to help the beleaguered crewman. They tethered the payload to a powerful four-wheeled snowmobile, revving it in reverse to keep the payload from swinging into the launch vehicle. Things were not looking good. Indeed, by 4:00 A.M. the launch chief had called it off.

A whole night had gone by. The scientists put their payload back in the hangar. The rest of us wrapped up and headed back to McMurdo for a nap. I woke up in time for lunch and wandered over to the galley. A couple of the crew were there, bleary-eyed from the night-long effort.

"The weatherman's sleeping, or hiding from a lynching," one of them joked.

"Now, wait a minute, he's the best one we've got," the other said.

"He's the *only* one we've got," the first replied.

There would be other opportunities for weatherman Ross Hays to nail the forecast during my visit. I was most interested in seeing the launch of BESS, an experiment capable of sifting through billions of cosmic-ray particles for traces of antimatter: antiprotons, in particular, and maybe even an antihelium nucleus. Even a single sighting of antihelium would shake up cosmology, for it would point toward vast regions of antimatter in space, entire stars and galaxies made of it. This would help shed light on a mystery that has dogged physics for decades. What happened to all the antimatter that should have been produced in the wake of the big bang?

On December 10, 1933, Paul Dirac, a thirty-one-year-old English physicist, accepted the Nobel Prize for some elegant mathematical work that added a whole new dimension to our understanding of the universe. Dirac had combined quantum mechanics and Einstein's equations of special relativity to describe electrons and their interactions with photons—a feat that until then had stumped the very brightest of minds. One of the strangest predictions of the so-called Dirac equation was the existence of a particle like the electron but with a positive charge: an antielectron, a *Doppelgänger*. A mere two years after he made the prediction, the positron was discovered, and Dirac's genius was firmly established. Antiparticles were now real: particles that are in every way like the ones we know but with the opposite charge. A positive electron, a negative proton, antineutrons—all adding up to the possibility of antinuclei and antiatoms.

In his Nobel lecture, Dirac painted a picture of an otherworldly universe that beggars belief even today:

> If we accept the view of complete symmetry between positive and negative electric charge so far as concerns the fundamental laws of Nature, we must regard it rather as an accident that the Earth (and presumably the whole solar system), contains a preponderance of negative electrons and positive protons. It is quite possible that for some of the stars it is the other way about, these stars being built up mainly of positrons and negative protons. In fact, there may be half the stars of each kind. The two kinds of stars would both show exactly the same spectra, and there would be no way of distinguishing them by present astronomical methods.

Once antimatter's existence was established, it became clear that nature should have had no preference for matter over antimatter: Both should have been produced in equal amounts after the big bang. Given that, they should have annihilated each other. Very obviously, that hasn't happened. Our universe, for all one can tell, is made entirely of matter. Where has all the antimatter gone? One solution is presented by theories that favor a slightly greater production of matter over antimatter in the early universe: There could have been an extra matter particle for every 10^{10} particles created. And when all other antimatter and matter had been mutually destroyed, the extra matter particles that remained would have formed this material universe. Such theories essentially break the symmetry between matter and antimatter.

But what if the symmetry between matter and antimatter was broken only in localized regions, leading to an excess of matter in some and an excess of antimatter in others? Then, as the universe inflated, these regions could have separated before the matter and antimatter particles could completely annihilate each other—thus some unannihilated regions of antimatter might still exist in our universe. As Dirac speculated, there could be entire stars and galaxies made of antimatter. One way of finding such regions would be to look for radiation in the form of gamma rays at specific energies, or "lines," coming from the annihilation of matter and antimatter at the boundaries. No such gamma-ray lines have been seen, indicating that the nearest concentration of antimatter must be at least 10 megaparsecs distant and so cannot be anywhere within or near our galactic neighborhood. (A parsec is a unit of length roughly equal to 19 trillion miles, or 3.26 light-years, and a megaparsec is a million times that; 10 megaparsecs would take you two-thirds of the way to the Virgo cluster, the nearest large galaxy cluster.) "But it is still a mystery, and you can't rule out the existence of antimatter regions," said John W. Mitchell, an affable astrophysicist from NASA's Goddard Space Flight Center, in Greenbelt, Maryland, and the U.S. principal investigator for BESS. As PI, Mitchell was leading the work, together with the PI for Japan, Akira Yamamoto of the High Energy Accelerator Research Organization (KEK) in Tsukuba.

One of BESS's tasks is to solve this mystery. To find out exactly how, I caught up with Mitchell inside the BESS payload building on the Ross Ice Shelf. It was a few days before the December 23rd launch, and Mitchell and his American and Japanese colleagues were taking care of last-minute tweaks to their equipment. The payload's shiny surfaces masked a huge superconducting magnet. Once BESS was aloft at an altitude of about 120,000 feet, cosmic-ray particles would smash into it, and the magnet's powerful field would deflect the particles. Negatively charged particles bend one way, positively charged particles bend the other. The higher the electric charge of the particle, the deeper its curve. The higher its momentum, the shallower its curve. BESS accurately measures the curved trajectories of the particles passing through it. This helps determine the nature of the cosmic ray.

The trouble is that an electron or a negative muon (a more massive cousin of the electron) can mimic an antiproton. At the same energy, they exhibit the same "magnetic rigidity"—that is, both will bend the same way and by the same amount—so other detectors surround the magnet to differentiate between such particles. Antiprotons are 1,836 times heavier than electrons and 8.88 times heavier than muons, and, for the same energy, the velocity of an electron or muon is much higher than that of an antiproton. In BESS, layers of plastic scintillators, which emit light as particles zip through them, measure their velocities with a precision of about a tenth of a billionth of a second, equivalent to the time it takes light in a vacuum to travel 3 centimeters. The scintillators also measure the electric charge of particles. Combine these measurements with magnetic rigidity, and they give you the mass of the particles and differentiate between antiprotons and the lighter background particles. But for very-high-energy particles, equivalent to antiprotons traveling at about 90 percent of the speed of light, even plastic scintillators are not good enough. Yet another detector, this one made of a silica gel, comes into play. Very-high-velocity particles, traveling faster than the speed of light in the gel, produce Cherenkov light as they pass through the gel, giving a measure of their velocity. With this trio of detectors, BESS is well equipped to look for antimatter.

The search for antimatter has become very sophisticated. It was not until 1979 that scientists, led by Robert Golden of New Mexico State University, were able to detect the first antiprotons from deep space, most of which are produced when primary cosmic-ray nuclei hit molecules of interstellar gas. Some by-products of these collisions are so-called secondary antiprotons, and they are created by known interactions in particle physics. But even these secondary deep-space antiprotons are rare, with one antiproton for about a hundred thousand protons, and they must be detected in a large background of electrons and muons. By about fifteen years ago, magnetic-rigidity spectrometers like BESS had improved to the point where antiprotons could be absolutely identified by their mass and unambiguously separated from the background, though the numbers detected were small.

The first-ever BESS flight, in 1993, saw six antiprotons. Another balloon-borne experiment a year before, called IMAX (Isotope Matter-Antimatter Experiment), had seen sixteen. Mitchell had worked on both with his Goddard colleague Robert Streitmatter. "Our joke was that we named every one of them, because we really looked at everything about them individually," he said.

BESS flew almost every year from 1993 to 2002, and starting in 1997 the experiment improved so much that it was seeing about five hundred antiprotons every day it was aloft. These secondary antiprotons are now so well understood that the new BESS is looking for an excess, a number higher than the expected secondary spectrum. One source could be antiprotons from the evaporation of small primordial black holes that formed just after the big bang. Another source could be the annihilation of dark-matter particles. Yet another source could be leftover regions of antimatter—but because antiprotons can be produced by processes in today's universe, finding an excess would not constitute absolute proof of the existence of primordial antimatter. An antinucleus heavier than a proton, on the other hand, almost certainly would, and this is what the BESS team is hoping to see. The probability that a collision of cosmic rays with interstellar gas would produce antimatter nuclei heavier than an antiproton is vanishingly small. So if BESS sees an antihelium (a nucleus with two protons), it would almost certainly have come from an antigalaxy—an entire galaxy made of antimatter. And just like the first few antiprotons, "an antihelium would get extremely personal," joked Mitchell. "We are unlikely to have more than one of those." But one antihelium would do. It would be stunning experimental proof that the visible universe contains primordial antimatter.

By the time this BESS flight is done with its data gathering and analysis, it will have detected almost 5 billion "events," including electrons, positrons, muons, protons, antiprotons, and atomic nuclei. The team then has to sift through this immense pile of particles to find a single antihelium. The analysis takes years. Preliminary results aren't expected until the summer of 2009 at the earliest. But Mitchell and others were in for the long haul. As he readied himself for the launch, he said, somewhat wistfully, "One absolutely undeniable, unquestionable antihelium would change the scope of physics forever."

Ballooning in Antarctica goes back to the heroic era of the continent's exploration (1895–1917). While Scott was beaten to the pole by Amundsen, he does hold another record. On his 1901–1904 *Discovery* expedition, Scott brought along two balloons and fifty cylinders of highly pressurized hydrogen gas. Each tethered balloon could lift a single observer. Captain Scott, as was his privilege, chose himself, "perhaps somewhat selfishly," to be the first man to rise over Antarctica. On February 4, 1902, as they were sailing toward McMurdo Sound, Scott stopped at a bay along the towering edge of the Ross Ice Shelf. The bay came to be called Balloon Bight (and was later renamed the Bay of Whales by Shackleton). The crew spread a large sailcloth on the snow, set the balloon on it, and connected it to the cylinders. Despite emptying the contents of sixteen cylinders into the balloon—the amount it was designed for, a total of 8,000 cubic feet—the balloon did not fully inflate, because the subzero temperatures had reduced the volume of the hydrogen. "It was not until we had brought out and emptied three other additional [cylinders] that its name 'Eva' could be read on a smooth, unwrinkled surface," wrote Scott.

Scott got into the basket and ascended to an altitude of about 800 feet, becoming the first person to see over the edge of the Ross Ice Shelf. Glaciers extended to the south, as far as the eye could see. (Today, it's known that the ice shelf extends nearly 650 miles into the heart of Antarctica, rising up from the coast in a series of undulations.) Next, Shackleton, who was serving under Scott on the expedition, went up in the balloon with a camera and became the first to take aerial photographs of the continent.

Scott and Shackleton would have been flabbergasted at the size and strength of the balloon that would launch BESS. Every piece of it dwarfed the one used by Scott. The balloon has a volume of 37 million cubic feet and uses helium instead of hydrogen. It soars into the stratosphere to a height of about 120,000 feet. The balloon's fabric alone weighs 5,300 pounds, and it carries a payload weighing nearly 6,000 pounds. From its top to the bottom of the payload, the balloon stands 1,000 feet tall, higher than Scott rose during his flight.

It was about midnight on the Ross Ice Shelf. The sun was behind a cloud. Everyone was preparing for yet another early morning launch. Mitchell stood near the launch vehicle admiring BESS, which was hanging from the crane boom. Thankfully, the sky was clearing, just as weatherman Ross Hays had predicted. "When we came out at five o' clock, it was bad," said Mitchell. "It was cold, it was windy, it was completely overcast, a dark dense cloud. Ross seems at this point to have it . . ." He trailed off, possibly not wanting to jinx the launch. The clouds were indeed clearing, and the summit of Mount Erebus came into view. Mitchell, a veteran of more than ten balloon launches, was far from jaded. "I'm going to scoot up to the top of the launch vehicle to take a picture," he said and sauntered off.

Meanwhile, Hays was busy monitoring the winds. Wondering how he was doing, I went into the rigging building. Anne Dal Vera, Austin's assistant (and a member of the 1992–1993 American Women´s Expedition to Antarctica, the first women to ski from the coast to the South Pole), was blowing up a meteorological pilot balloon, or "pie ball," for the hourly updates on wind speeds. "I feel a bit like a little kid, walking out there with my big red balloon," she said, as she adjusted the amount of helium so that the balloon hovered in place when attached to a standard weight. "I'm ready now," she said to Hays.

"Ready? On three . . . One, two, three," shouted Hays.

Red balloon in hand, Dal Vera bounded out of the rigging room to a platform outside. Exactly sixty seconds later, she removed the standard weight, released the balloon, and started tracking it with a theodolite—a small sighting telescope—connected to Hays's computer. They knew the exact rate of ascent for the balloon, so tracking its position every thirty seconds gave them enough data to calculate wind speed and direction at various altitudes. "She's tracking now," said Hays. "The first measurements are coming in."

Based on the pie balls, Hays could predict with some confidence when the conditions would be amenable for launch. They were not there yet. Just then, Austin walked in and pulled me aside. Erich Klein, the launch manager, had requested that I leave Hays alone so that he could concentrate on his weather forecasting. Feeling a bit chastised, I walked over to the galley for a coffee. Minutes later, Klein walked in to chat. He was clearly under pressure. Of the three balloon payloads to be launched this season, they had managed only one. Christmas was approaching, and there was little time left. "We need all these off by the first of January," he said. "The third pay-load—he's got the hind tit, as we say at home." Home was Palestine, Texas, where NASA's Columbia Science Balloon Facility is situated. "We are at the mercy of the weather," he said with a shrug.

Indeed they were. And the news was not so good. After a promising night, the weather was refusing to cooperate yet again. The crew, who had been on the ice for hours, started filtering back into the galley for coffee, cookies, and hot food. And commiseration. Even the heavy-vehicle crew came in, including a bear of a Texan called Reid Chambers. During the launch, he would be responsible for standing on the launch vehicle and keeping the 2-ton payload from swinging around. The bearded Chambers, with his brown, padded, extreme-cold-weather gear and his deliberate movements, looked something like a giant-size Ewok from *Star Wars*. In a voice that was strangely soft for a man of his stature, he addressed those of us who had come to watch the balloon launch: "You're addicted now, aren't you? You've seen one launch and you have to see another. Balloon groupies. It happens." Then to everyone else who wasn't working on the launch, he added, "It's three A.M. Y'all know you don't have to be here. It's nearly Christmas."

"For what it's worth," said Austin.

"There's two more shopping days left," quipped Chambers.

A crew member in full ECW gear, dark glasses, and a hood covering his face walked into the galley and shouted, "This is a holdup!"

"Take my wife," someone piped up.

The back-and-forth continued until Chambers muttered a somber reminder of their predicament. "Sometime in the next eighteen to twenty-four hours, I'll get to sleep again," he said.

Klein was asked whether he was calling off the launch. "It's still early. We haven't even thought of giving up," he said. "It's looking good, except for the first two layers." The winds nearer the surface were not slowing down.

At 4:00 A.M., Mitchell walked over and helped himself to some food. "I'm tempting Murphy's Law," he laughed. "Is it worse for Murphy's Law to stop me from eating or to stop the launch?" As it happened, within minutes we got news that the weatherman had cleared the launch. The winds would die down in a couple of hours. Preparations needed to begin now.

The moment the weatherman gave the word, the heavy-vehicle crew swung into action. All the equipment—which had been brought by cargo ship to McMurdo after icebreakers had cleared the way—had to be assembled for launch. A Caterpillar tractor crunched toward the launch pad, hauling a box with a neatly folded balloon. Two other tractors each pulled trailers loaded with twelve giant tanks of helium. The launch vehicle was already on the ice, parked nearly at the center of the circular launch pad, its boom aligned to the predicted direction of launch-time winds. The payload hung from the boom. The crew connected the payload to the parachute and set it down behind the launch vehicle. At this point, they could still be called back if the winds changed. But the conditions continued to look good, so the crew started assembling the balloon. A long strip of canvas was laid over the snow between the box containing the balloon and the parachute. Gingerly, the crew pulled out the folded balloon, sheathed in bright red plastic, and set it out on the canvas. Even the immensity of the ice shelf could not dwarf its size, as it stretched from one end of the launch pad, where the helium tanks stood, to the payload and parachute near the center. The work was done with extreme care, for the parachute is made of polyethylene plastic film a mere

0.0008 inch thick—about the thickness of a sandwich wrap. Then the crew cabled the balloon to the parachute and wired up its control electronics, working quickly before the cold rendered their fingers useless.

It was time to inflate the balloon. There was no going back now. A section of the balloon was fed under a large spool, which was mounted on a sled being pulled by a Caterpillar tractor. Two tentacle-like fill tubes coming out of the sides of the balloon were connected to valves, each held by a crew member. The valves were attached to long, thin, black hoses that snaked over the snow to the helium tanks. Everyone was asked to move away, because a broken hose, with pressurized helium flowing, can lash out with enough force to sever a leg.

The helium was released from the tanks, and the fill tubes engorged, looking like transparent billowing beasts that had to be tamed by the crew. The high-pitched whine of the pressurized helium rent the air. The top of the balloon filled up and rose. Dragging the spool sled slowly toward the payload, the crew let more of the balloon through, until all of the helium had been injected. The fill tubes were disconnected and tied up tight. Filling the balloon had taken nearly an hour, and canceling the launch now meant the balloon would be lost. The expensive helium, too. Luckily, the winds were fine. I heard Klein, the launch manager, shout, "Release the balloon, release the balloon!" The directive rang out on all hand-held radios around the launch pad. The crew disengaged the spool, which swung quickly aside. The freed balloon rose in slow motion, with a mighty flapping of plastic. As Scott had discovered, the Antarctic cold reduces the volume of a gas. As a result, the balloon didn't fully inflate. It looked like a giant translucent jellyfish, its tied-up tentacle-like fill tubes rounding out the picture.

In a matter of seconds, the balloon was above the launch vehicle. If the wind is calm, the balloon should hold steady right over the payload, but often that's not the case. The balloon blows with the wind, however slight, and it can drag the launch vehicle around with it, so the driver has to start maneuvering to keep the payload directly beneath the balloon. But the balloon is directly above the vehicle, so for all practical purposes he's driving blind.

To help out, the crew chief was standing on the launch vehicle, looking up at the balloon and shouting directions into his radio: Left, right, forward, back, slow down, speed up . . . until the vehicle was moving at just the right speed and direction and the balloon and the payload were vertically lined up. It was time to let the balloon go. The driver braked and reversed direction. The crew pulled a pin that connected the payload to the boom, and the balloon took up the weight of the payload. This is always a tense moment, for even a tiny error can cause the payload to swing like a pendulum and smash into the vehicle or into the ground.

If it goes right, though, it's transcendent.

The heavy payload hovered above the ground for a moment, then started to soar. Five hundred feet per minute. Everyone on the launch pad cheered, some for joy at having seen something magical, others at the relief of a launch perfectly executed. Then a hush descended as the balloon rose against the backdrop of Mount Erebus. It was still long and thin and far from its final shape. The payload swiveled beneath, its shiny surfaces and solar panels glinting occasionally as they caught the sun. The balloon looked fragile, the payload even more so. For a while, the whole ensemble seemed to slow down, partly because the cloudless blue sky offered no sense of scale, but also because you perceive that something is rising only if it is getting smaller. In this case, the balloon was getting larger by the minute. The helium was expanding as the atmospheric pressure dropped. Everything seemed suspended, allowing us below to catch our breath and marvel.

Within about three hours, the balloon would punch through the troposphere, after experiencing temperatures as low as −49°F. The helium would expand to more than two hundred times its volume during launch, and the balloon would be fully stretched, resembling a giant diaphanous pumpkin nearly 400 feet across. Against a blue sky, it would be almost half the size of the full moon. Airline pilots flying below such launches, unable to discern the size of this curious object, have repeatedly been tricked into thinking that the balloon is closer to them than it actually is—at 120,000 feet, however, it's twice as high above the airplane as the plane is above the ground. Hours later, I could see the balloon from McMurdo Station, as it flew over the Transantarctic Mountains on its long journey around the South Pole.

The South Pole was where I was headed next, to see the most ambitious detector ever conceived: the IceCube neutrino telescope, which is hoping to see neutrinos from outer space smash into a cubic kilometer of clear Antarctic ice. Only in Antarctica is such a telescope possible, for nowhere else can you find such a mass of ice so stable and solid.

The weather during the early hours of the day before I left McMurdo was cold and crisp. Keen to partake of a clear day after two days of storms, I headed toward Observation Hill, or Ob Hill, a volcanic cone that rises steeply to a height of almost 800 feet. After walking on gravel trails, I found the final few hundred feet to be more of a scramble on rocks. From the top, McMurdo's sprawl was impressive. Giant white fuel tanks in the foreground dominated the scene. Bright-colored shipping containers provided relief against a relentless black, brown, and white background. Red helicopters with blue tails stood on the helipad, looking like Lego toys. Beyond the station was the Hut Point peninsula, and at its tip stood Scott's Discovery Hut. The sea ice around the shores of the station was getting thin; melt pools gleamed in the sun. The sea-ice runway, abandoned for the season due to the thinning ice, stood out clearly, stripped of the snow that covered the rest of the frozen sound.

To the south-southeast, toward the pole, was the Ross Ice Shelf, the Barrier that had so dominated the imagination and the lives of the early Antarctic explorers. From here, it was easy to see why. There wasn't an end in sight, even on this clear day; the white ice merged into the powder-blue sky. At the summit of Observation Hill, the only sound was the hum of McMurdo's generator. And the wind. All else was quiet, as if hushed by the weight of the ice. Tibetan prayer

flags at the summit fluttered in the stiff breeze—a recent addition to
what was already a symbolically religious place. On a wooden cross
about eleven feet high were etched the names of five explorers: Capt.
R. F. Scott, Dr. E. A. Wilson, Capt. L.E.G. Oates, Lt. H. R. Bowers,
and Petty Officer E. Evans, "who died on their return from the pole
March 1912." Below it was the last line of Tennyson's "Ulysses": "To
strive, to seek, to find, and not to yield."

When Scott and his men did not return from the South Pole by
the summer of 1912, expedition member Cherry-Garrard and others went in search of them. On November 12th, they found the frozen bodies of Scott, Wilson, and Bowers in their tent. Their note-
books revealed that Evans had died in February near the bottom of
the Beardmore Glacier and that Oates, unable to travel any longer,
had died a month later, having voluntarily walked into a blizzard.
The search party erected a funeral cairn and left the bodies inside it.
The ice was to be their final resting place. Then they trekked back to
Ross Island and waited for the return of their ship, the *Terra Nova*,
which had left McMurdo Sound for the winter. On January 18, 1913,
the ship appeared from behind a glacier and hailed them: "Are you all
well?" The men on land said yes and then shouted out the bad news:
"The Polar Party died on their return from the Pole. We have their
records."

The survivors decided to erect a cross in memory of Scott's team.
Cherry-Garrard wrote later: "Observation Hill was clearly the place
for it, it knew them all so well. Three of them [had] lived three years
under its shadow: they had seen it time after time as they came back
from hard journeys on the Barrier . . . It commanded McMurdo Sound
on one side, where they had lived: and the Barrier on the other, where
they had died. No more fitting pedestal . . . could have been found."

Writing ten years after the tragedy, Cherry-Garrard foresaw bases
like McMurdo and the South Pole station:

> I do not suppose that in these days of aviation the next visit to
> the Pole will be made by men on foot dragging sledges, or by
> men on sledges dragged by dogs, mules or ponies; nor will de-

pots be laid in that way . . . I hope that by the time Scott comes home—for he is coming home: the Barrier is moving, and not a trace of our funeral cairn was found by Shackleton's men in 1916—the hardships that wasted his life will be only a horror of the past, and his *via dolorosa* a highway as practicable as Piccadilly.

While the route from McMurdo to the South Pole is hardly as practicable as Piccadilly—one of central London's busiest streets—neither is it nearly as dangerous as when Shackleton, Amundsen, and Scott made their journeys. The U.S. Antarctic Program makes some 350 flights from McMurdo to the South Pole during the hundred days of clement weather each year, mainly in support of extreme science. At the South Pole, theory is more than meeting its experimental match. Deep within the polar ice sheet, a neutrino telescope is on its way to becoming one of the only instruments that can probe the effects of quantum gravity. All I had to do was wait for my name to show up on the flight manifest.

8 · EINSTEIN MEETS QUANTUM PHYSICS AT THE SOUTH POLE

N THE ICECUBE LOUNGE, burly, bearded men were talking murder. How easy would it be to dispose of a dead body at the South Pole? All you'd have to do is dump it down one of the holes being drilled for the IceCube neutrino telescope. Each hole is kilometers deep and reaches almost all the way down to the bottom of the East Antarctic Ice Sheet. The body would be impossible to find.

The burly, bearded men were drillers. The IceCube Lounge is an informal corner of the game room in the new U.S. Amundsen-Scott South Pole Station, where hard-drinking men sit down each evening and let off steam after a grueling nine-hour shift. They had just finished drilling one of the holes they were now joking about. The hole was deep enough to stack eight Eiffel Towers in. After observing the drillers do their bit, I had intended to stay and watch the next team start lowering a string of bulbous light detectors into the ice, but it was late and I had to get back to the station. The temperature was −40°F, if you counted the wind chill, and was biting through my shoes and two layers of gloves. My head felt heavy and congested, an effect of the altitude, which is 9,200 feet above sea level. I had flown in from McMurdo just two days earlier and had yet to fully acclimatize.

The flight from McMurdo was surprisingly short. The plane, a ski-equipped LC-130, flew over the Ross Ice Shelf and past the massive Beardmore Glacier—the same glacier that Ernest Shackleton in 1908 and then Robert Scott in 1911 had scaled during their ascent over the Transantarctic Mountains on their way to the South Pole. They had dragged their sledges to the top, tackling the swarms of crevasses that form as the glacier glissades down from the heights of the East Antarctic Ice Sheet. Shackleton, who discovered this route, wrote in his diary, "It is a wonderful sight to look down over the glacier from the great altitude we are at, and to see the mountains stretching away east and west . . . Thank God, we are fit and well and have had no accident, which is a mercy, seeing that we have covered over 130 miles of crevassed ice."

After climbing over the Beardmore Glacier, Shackleton and three other members of his *Nimrod* expedition came within 97 miles of the South Pole, but they turned back, due to a shortage of rations. Shackleton's unselfish act—for he could have ordered a suicidal march to the pole—earned him considerable acclaim and no less than a knighthood. Raymond Priestley, a member both of Shackleton's *Nimrod* expedition and Scott's subsequent *Terra Nova* expedition (during which Scott, in his attempt to become the first to reach the South Pole, was undone by the Norwegian Roald Amundsen), said this about the pioneers of Antarctic exploration: "Scott for scientific method, Amundsen for speed and efficiency, but when disaster strikes and all hope is gone, get down on your knees and pray for Shackleton."

Amundsen's reputation for speed and efficiency was sealed when he raced Scott to the pole during the summer of 1911–12. Scott followed Shackleton's trail, while Amundsen took a shorter route over the Transantarctic range, choosing to ascend the steeper Axel Heiberg Glacier, which falls in icy cascades between two imposing mountains. Later, the Norwegian explorer Fridtjof Nansen wrote about his countryman's courageous choice to forgo the Beardmore Glacier, one of the factors that helped Amundsen beat Scott to the Pole. Nansen wrote, "People thought it a matter of course that he would make for Beardmore Glacier . . . We who knew Amundsen thought it would be more like him to avoid a place for the very reason that it had been trodden by others. Happily we were right. Not at any point does his route touch that of the Englishmen — except by the Pole itself."

Halfway through the flight, I squinted out of the tiny porthole next to me. We had already crossed the continent's spine — the Transantarctic Mountains — and were flying over the vast whiteness of the Antarctic Plateau. All I saw were nunataks — the dark-brown tops of the tallest peaks — the bulk of the mountains being submerged beneath a mile or two of ice. I dozed off. When I woke up, a crew member was frantically signaling with his hands to get our attention (we were all wearing earplugs, which drowned out everything except the muffled drone of the airplane's engines). But his gesture was clear: "Strap up, we are landing." I looked at my watch. Barely three hours had passed since we left McMurdo. As the plane skied to a landing on the ice at the South Pole, I was surprised by the sight framed in the porthole: a place strewn with freight containers, Jamesway huts, timber stacks, Caterpillar tractors, cranes, and Ski-doos. It could have been a construction site somewhere in the United States, albeit in deep winter.

That thought was dispelled as soon as we stepped out of the plane. The South Pole in summer was colder than any winter I had ever experienced. A staffer quickly shepherded us toward the station, just a few hundred yards from where the plane had stopped. A handful of people were waiting to board the plane, which was heading back to McMurdo and had never cut its engines. Doing so in the extreme cold would cause the motors to freeze, and enormous effort would be required to get them going again. The McMurdo-bound passengers had thirty minutes to get on board, for that's about the longest the pilots like to stay parked on the ice. Any longer and they would be wasting precious fuel — or worse, the skis could ice over and get stuck.

If the cold wasn't telling enough, the short walk from the plane to the station made it clear that we were at the pole. By the time I had climbed the first flight of stairs leading into the building, weighed down by my extreme-cold-weather gear, I was winded. The pole's 9,200-foot altitude feels a lot higher. The oxygen pressure is lower here than at similar altitudes elsewhere, so the body has to work extra hard to get oxygen from the lungs into the bloodstream. My chest hurt. I felt as if I had just run a mile.

We were welcomed by Beth Watson, the station-support supervisor, who gave us a quick briefing. She cautioned us that water was an extremely scarce resource and asked us to limit ourselves to two 2-minute showers per week. It was an odd request, considering that we were surrounded by water—except that it was frozen, and thawing ice at these temperatures takes immense energy. However, there were no limits on drinking water; in fact, we were asked to drink copious amounts to stave off dehydration, which is a serious hazard here. It was important to watch for any signs of altitude sickness: a pounding heart and shortness of breath despite resting after a walk or a climb, a splitting headache, even persistent light-headedness. Also, she asked us to maintain a respectful silence in our rooms; the scientists and staff at the pole were working around the clock in three shifts, so there were always people sleeping in the berths. Being quiet, then, was a rule, not a request. Finally, the food: It was some of the most expensive in the world, as every bit of it had to be flown in. "You take what you can eat and eat what you take" was the dictum. Welcome to the South Pole.

The South Pole remains one of the most remote places on Earth. The station is dependent on flights from McMurdo Station, not just for food but for everything else, from fuel to heavy machinery, and flights can be grounded by the notoriously fickle weather. Serious medical problems require evacuation to New Zealand, a considerable effort during the summer and a near impossibility in the dark and extreme cold of the polar winter. "You can't really ever get complacent when you work here," said South Pole area director Bettie Grant, whom everyone called BK. "As soon as you do, this place will reach out and kick your butt."

As if to reinforce just how difficult things can get, BK had a framed photograph outside her office. It was taken in April 2001 and showed a Twin Otter silhouetted against bright lights that blazed in the winter darkness. The plane had flown all the way from Calgary, Canada, to evacuate the station's doctor, Ronald Shemenski, who had been diagnosed with gallstones and urgently needed surgery. Initially, the U.S. Air National Guard had mobilized three LC-130 aircraft to medevac Shemenski, but everyone soon realized that an LC-130 would not be able to handle the conditions—mainly because its hydraulic landing gear could freeze in temperatures that were touching –100°F. The LC-130s, which had already flown from New York to Hawaii, en route to Christchurch, were sent back, and an alternate plan was initiated. Two Twin Otters started from Calgary and stopped in Chile and then at Rothera, the British Antarctic Survey's base on the Antarctic Peninsula. One plane stayed behind at Rothera as backup, while the other flew 1,346 miles across the Antarctic Plateau—a 10-hour flight—to reach the South Pole. The plane's simpler controls and fixed skis allowed it to land safely. The temperature at the time was –90°F. The plane, which had flown in a replacement doctor, medevacked Shemenski to Chile. From there he flew by commercial airliner to Denver, Colorado, where a routine EKG prior to surgery showed that he had heart problems. An angiogram revealed blockages in two arteries. The doctors immediately opened the blocked arteries and implanted stents. Shemenski went home to Ohio to rest before returning to Denver to have his gallstones removed. It wasn't until the end of June 2001 that he was able to rest easy.

My first act after attending the briefing was to walk over to the geographical South Pole marker, where I found a large wooden plaque honoring both Amundsen and Scott. Just what it meant to Amundsen to be the first man to reach the South Pole is evident in what he said when he passed the southernmost point reached by Shackleton in 1909. Amundsen wrote, "No other moment of the whole trip affected me like this. The tears forced their way to my eyes; by no effort of will could I keep them back." On the wooden plaque are his rather more prosaic words, recalling his history-making achievement on December 14, 1911: "So we arrived and were able to plant our flag at the geographical South Pole."

I stood there imagining a South Pole without flags, buildings, or people—just an expanse of snow. It wasn't hard to do. While the pole is crowded today, the populated region is but a speck. The snow-covered ice sheet, nearly 3 kilometers thick at the pole, stretches from horizon to horizon. For Amundsen and his men, the first people to set foot here, there would have been nothing except ice all around. One can almost visualize the curvature of the Earth here, such is the unbroken expanse.

The South Pole was no longer virgin territory when Scott arrived on January 17, 1912. On the same plaque, another quote recalled his anguish: "The Pole. Yes, but under very different circumstances from those expected." In little more than a month, the same place, pristine but for remains of a camp left behind by Amundsen, had a devastating effect on Scott. Words that appear not on the plaque but in his diary reveal the true extent of his torment: "Great God! This is an awful place and terrible enough for us to have laboured to it without reward of priority."

The awfulness was not apparent to me as I stood and gazed at the new U.S. South Pole station, which replaced a geodesic dome dating from 1975 and was still being completed. It is a sleek two-storey block, clad in gunmetal-gray siding and chock-full of comforts. In fact, once inside the new station, it's easy to forget just how far from civilization you are. People walk around in T-shirts and jeans, even shorts and flip-flops. The chefs in the galley serve breaded catfish, Chinese pork roast, and New York strip steak. Noticeboards carry messages advertising yoga sessions and a pub trivia game in the galley. Walk down the corridors that connect the station's two pods and you pass a clinic, a library, a game room, a hydroponic greenhouse, a post office, and a science lab, as well as a gym with treadmills and exercise bikes overlooking a basketball court. The pods have private rooms for 150 residents, all with phone and Internet connections, thanks to a satellite link, which is available for about ten hours a day.

One scientist told me that the new station reminds him of a cruise ship, and it's not a bad analogy. From the windows that line the galley, the view across the snow-covered plateau stretches all the way to the horizon. When the sun catches the sastrugi—wind-whipped ridges in the snow—they glint and cast shadows, like whitecaps on a turbulent sea.

But step outside and you are reminded that everything at the South Pole is beholden to the snow. The residents fight a Sisyphean battle against the relentless drifts, which pile up against any conventional structure and eventually bury it. The first station, built in 1957, is now more than 20 feet below the surface. A geodesic dome 50 feet high and 160 feet across was built some fifteen years later and sheltered several buildings inside it. For decades, the dome was the iconic structure of the South Pole, but it, too, has succumbed. The new station is different: It has an aerodynamic structure and is built on giant pillars. The streamlined profile is designed to funnel the drifting snow underneath the building and out the other side, preventing it from piling up on the windward side. But that just delays the inevitable. Eventually, the snow will stack up. Then the building will be jacked up on the giant pillars. This can be done once, possibly extending the structure's lifetime to a total of about forty years. But only time will tell how long the new station can fight off the snow.

Though most of the people here are scientists and support staff, it is mainly geopolitics that has kept the United States entrenched at the pole. As the old dome struggled with the snow and an increasing population, a National Security Council memo dated March 9, 1996, made it clear that a continued presence at the South Pole was considered vital: "Abandonment of the station would create a vacuum and likely result in a scramble to occupy the site, to the detriment of our position as well as to the stability of the Treaty system." This is the Antarctic Treaty, which preserves the continent for peaceful pursuits and suspends all territorial claims. While the United States itself has never made any formal claims, the South Pole station gives it a foothold within the territories claimed by other countries, all of which originate at the pole. If the treaty were to lapse, the United States would be well placed to challenge any aggressors. This has meant that the U.S. National Science Foundation (NSF) continues to fund science at the pole even as American research programs elsewhere suffer, and for the first time the science budget dwarfs funding for the station itself. While the new station cost $153 million, the IceCube neutrino telescope alone, which is being built and operated by an international consortium led by the University of Wisconsin and is funded mostly by NSF, will cost more than $270 million. But many consider that a small price to pay to detect neutrinos from deep space. The payoff would be even bigger if IceCube manages to successfully detect effects of quantum gravity, perhaps the most intractable problem in physics today.

The conflict between general relativity and quantum mechanics has flummoxed physicists for decades. The former is a theory of gravity on the grandest of scales—that of the universe itself. Quantum mechanics, on the other hand, illuminates the behavior of the smallest particles. Both work amazingly well at explaining the observed world, except in situations where they need to work together, such as inside a black hole, where enormous gravitational forces have to contend with matter compressed into microscopic volumes. Or in the very early universe, when spacetime, which is described by general relativity, was so small that the rules of quantum mechanics also held sway. Quantum mechanics has been successfully used to describe three of the four fundamental forces of nature: the electromagnetic force and the weak and strong nuclear forces. For each of these forces, physicists have what are called field theories—descriptions of fields specific to each force. Each force, in turn, is mediated by the exchange of particular particles—photons for the electromagnetic force, W and Z bosons for the weak nuclear force, and gluons for the strong nuclear force. The odd one out is gravity. There is no field theory for gravity, no particle that we know of that is exchanged between entities feeling its force (other than the hypothetical graviton). This is profoundly unsettling to physicists, who think that all the fundamental forces were unified as one in the early universe and hence should all be described using the same physics.

To many, this incompatibility is a sign of a deeper underlying problem with either general relativity or quantum mechanics or both. The search for a unified theory of quantum gravity has proceeded on many fronts, the most promising being string theory, which has been extraordinarily successful in showing how all the fundamental forces, including gravity, can arise from the vibrations of microscopic strings. As we saw in chapter 6, these strings are thought to vibrate in ten dimensions of spacetime, six of which are so small as to be undetectable. Not surprisingly, there has been much strident criticism of string theory for being unable to make testable predictions. Another theory, called loop quantum gravity (LQG), says that spacetime, when you zoom in to the smallest possible scale, breaks up into a tangle of loops. In LQG, spacetime is no longer the smooth, indivisible fabric of Einstein's general relativity. It therefore allows spacetime and gravity to be described using principles of quantum mechanics. But LQG, like string theory, is a work in progress. It's difficult, given current technology, to conceive of experiments that can probe spacetime and particles at the scales required to test either of these theories.

That hasn't stopped progress altogether, however. Phenomenologists, who are concerned with taking empirical observations and relating them to fundamental theories, have started asking different questions. Are these approaches to quantum gravity telling us anything about the nature of spacetime—however vague—that we could test for? If any such quantum-gravity effects could be detected, it would help constrain the theories. It's important to assert that neither string theory nor LQG are anywhere close to making precise testable predictions of quantum-gravity effects. Any predictions would depend on the precise models that are built using these theories, and no one knows which, if any, of these models is the right one. "When I contemplate the sort of phenomena that might come from a quantum-gravity model, I don't feel obligated to look to specific worked-out examples of models for what they predict," said Dan

Hooper of Fermilab, near Chicago, Illinois. "I consider the work done in quantum gravity so far to be so unpredictive that I, as a phenomenologist, feel completely free to imagine it in almost any form."

The key, then, is to pay attention to hints from these theories about how nature could be different at scales at which gravity is quantized. One hint is that spacetime may not be smooth and infinitely divisible, as described in general relativity. Instead, at the smallest scale—of about 10^{-35} meters, which is called the Planck length—spacetime could be discrete, or possibly a roiling mess, a frothy concoction of black holes and virtual particles that pop in and out of existence. The other is the breaking of something called Lorentz symmetry, which is the cornerstone of special relativity. It says that the limiting velocity for any particle is the speed of light in a vacuum—that nothing can travel faster. If Lorentz symmetry were to be broken, then it's possible for different particles to have different limiting velocities: Electrons might have a different maximum velocity than, say, photons or neutrinos. It turns out that if spacetime is indivisible below a certain length scale, then it can lead to a violation of Lorentz symmetry. Could either the foamy or the discrete nature of spacetime or the violation of Lorentz symmetry or all three somehow show up in experiments? Normally, these effects, even if they exist, are too small to detect. But when particles travel across cosmological distances, these effects might add up to something measurable. And the particle most likely to be influenced by quantum gravity is the neutrino. IceCube could have something to say about all this, thanks in no small measure to the drillers.

It was also thanks to the drillers that I learned of the dangers of hot-water drilling at the bottom of the world. I first walked into the IceCube Lounge with some trepidation. Past the Ping-Pong and pool tables were a couple of sofas. Numerous bottles of beer and an oversized bottle of Kessler whiskey testified to an evening in progress. Jim Haugen, an engineer from the University of Wisconsin in Madison who was managing the IceCube operations, knew I was coming. He welcomed me and did a quick round of introductions. Dennis Duling, a brawny man with a generous beard, loomed large at the table (someone described him as "crotchety," and he grudgingly conceded that this was his normal disposition). The bottle of Kessler was his. Matthew "Red" Mathiason looked like he had just come in from the sun. Eric "Bear" Coplin—another bearded man of ample proportions—seemed too gentle for his nickname. The leader of the drill team was Alan Elcheikh, a laconic Aussie of Lebanese descent and a veteran of more than a decade of drilling in Antarctica. He was just Al. Discussing the IceCube drilling operation, he said, "I don't think anybody else has anything approaching this. It's monstrous."

Four thousand eight hundred eyes. Holes 2.5 kilometers deep. A cubic kilometer of clear ice. Since the austral summer of 2004–05, the IceCube team has been drilling in the East Antarctic Ice Sheet, aiming for a total of eighty holes, fifty-nine of which had been completed by March 2009. Into each 0.5-meter-wide hole is lowered something akin to a giant string of pearls. Each string is a kilometer-long chain of sixty digital optical modules (DOMs), designed to look for flashes of light emitted when neutrinos smash into the ice.

IceCube will be unlike any other neutrino detector for some time to come. The other advanced detectors, such as the one at Lake Baikal and those being built in the Mediterranean, are small by comparison. Just four of IceCube's eighty strings has more DOMs—digital versions of the analog photomultiplier tubes (PMTs)—than the entire complement of 228 PMTs under Lake Baikal. Though the Baikal detector was the first to use a natural body of water to detect atmospheric neutrinos and continues to study them while hoping to see neutrinos from distant astrophysical objects, IceCube will far surpass those capabilities, venturing into areas of physics that other detectors can only dream about.

Conceptually, IceCube is similar to the Northern Hemisphere's underwater detectors except that a neutrino at the South Pole smashes into ice instead of water. This collision creates a muon, which then speeds through the ice, forming in its wake a streak of light—a Cherenkov cone. The trick, as we saw in chapter 3, is to distinguish this sort of muon from those constantly pouring in from the atmosphere. It's an impossible task if you are looking at downward-going muons; a neutrino-induced muon will be outnumbered a billion to one by cosmic-ray muons. As with the Lake Baikal detector, IceCube looks for muons that are headed upward. Only neutrinos that have passed through the Earth and strike the ice from below can create such muons; none of the cosmic-ray muons from the other side of the planet make it through. Despite this common approach, the size and sensitivity of IceCube sets it apart, making it a telescope in the truest sense of the word.

All the DOMs in IceCube are synchronized with extraordinary precision by exchanging signals with each other and with a precise clock at the surface. Every so often, an LED (light-emitting diode) nearly 2.2 kilometers inside the ice sheet is flashed for a nanosecond, which the DOMs detect. This information is used to determine the position of each DOM and the geometry of the 3-D lattice of DOMs in the ice. Once synchronized and mapped, the DOMs look for light from muons, and their ability to pin down the arrival time of muons to within nanoseconds, combined with the gigantic size of the detector, allows IceCube to re-create muon tracks so accurately that it can discern where the original neutrino came from to within a 1° patch in the sky. That's like pinpointing a neutrino source in an area of the sky about as wide as two full moons. It's an astonishing capability, and it's what makes IceCube a telescope, not just a detector.

IceCube hopes to detect extremely energetic astrophysical neutrinos from sources such as active galactic nuclei, which are powered by enormous black holes that spew out jets of particles. Although these celestial objects also emit high-energy gamma rays and cosmic rays, it is the neutrinos that offer the best chance of understanding such high-energy astrophysical processes. That's because neutrinos flit effortlessly through virtually everything in their path, while high-energy gamma rays are absorbed by the cosmic microwave background and so never reach us. As for cosmic rays, the majority are deflected by magnetic fields as they travel, so their incoming direction rarely reveals their actual point of origin. "Neutrinos are the only particles with which we can look at high energies very deeply into the universe," said Wisconsin physicist Albrecht Karle, IceCube's associate director of science and instrumentation, as we sat in the surreal comfort of the IceCube Lounge.

But my interest in IceCube lay in its implications for cosmology. Like the Lake Baikal neutrino detector, IceCube will be looking for neutrinos emanating from the annihilation of dark-matter particles in our galactic center. However, IceCube has an unmatched potential for searching for new physics. When it's completed, IceCube could become one of the only instruments in the foreseeable future capable of indirectly testing the theories of quantum gravity, which suggest that at the Planck scale the fabric of spacetime is far from smooth. If neutrinos traveling from some distant astrophysical source, such as a gamma-ray burst, have wavelengths that are close to the Planck scale, they are expected to interact with this graininess of spacetime, and the IceCube neutrino telescope could be sensitive enough to see the effects of any such interactions. These neutrinos are much higher in energy than those being detected by other detectors, including the one at Lake Baikal. Far fewer high-energy neutrinos are created than lower-energy ones, so larger and larger volumes of water or ice are needed to catch them. Still, contemplating a neutrino detector that requires a cubic kilometer of clear Antarctic ice took some chutzpah.

It was a point that seemed to elude some members of a U.S. congressional delegation: distinguished visitors—or DVs, as they are called at the pole. The day I watched the drillers in action was the day the DVs flew in from McMurdo for a whirlwind tour of the South Pole station. I knew my timing was bad when I came to the site and found all the hardhats taken. The self-confessed "crotchety" Dennis Duling, his beard encrusted with snow, saw me staring into an empty bin usually overflowing with them. We walked over to a makeshift lounge fashioned out of a shipping container and found a hardhat. "Now you look like a driller," he quipped.

The drill camp was essentially a bunch of bright-red shipping containers that made up the offices and the hot-water plant, about a mile away from the South Pole station. To get to it, I had to walk past the skiway, making sure that the red warning beacon wasn't signaling that a plane was about to land. I crossed the CMB alley—a line of pioneering radio telescopes that take advantage of the thin, dry, stable atmosphere at the pole to study the cosmic microwave background. At the far end of the alley is the spanking new South Pole Telescope, which, with its 10-meter primary mirror, is so sensitive that it will use the CMB as a backlight shining at us from about 370,000 years after the big bang, against which it aims to pick up the "shadows" of the first galaxy clusters that ever formed.

Past the South Pole Telescope was the IceCube drill camp. White steam billowed from the chimneys atop the containers that housed the heaters. Just three colors—all of startling purity—dominated the scene: the red containers, the white steam and snow, and the deep-blue sky. Beyond the drill camp was the tower operating structure (TOS), where the drillers worked. Fat pythonlike insulated hoses snaked over the ice sheet toward the TOS, and in places where the insulation was worn out, the hot hoses had melted troughs in the snow. We walked over to the TOS, where the drilling was about to start. Two Swedish drillers and Red Mathiason were staring down a 40-meter-deep hole that had already been made in the firn, the layer of dense snow that lies on top of rock-hard ice. The hard part was to follow—drilling a 2.5-kilometer-deep hole into the ice sheet. The hot-water drill was already in the hole, ready to go. Steam was spewing out of it. The drill is essentially a narrow nozzle connected to a fat hose, through which near-boiling water is blasted out at a thunderous pressure of about 1,000 PSI. Plainly put, they were melting a huge hole in the ice.

While it looks deceptively simple, the drilling operation is actually a nightmare, causing Francis Halzen, the principal investigator for IceCube at the University of Wisconsin, Madison, no end of grief. ("If you ask me what I am thinking when I lie awake at night, it is not about neutrinos, it is [the drilling].") To start with, accidents are a serious worry. One occurred during IceCube's first season of operation. A Swedish driller was thrown into the air as he tried to grab a heavy cable that had started sliding down an open hole. He suffered a punctured lung and broken ribs, requiring emergency surgery at the station's basic medical clinic and immediate evacuation to New Zealand.

But more than accidents, it's the method of drilling that's the greatest cause for concern. The team needed one that would be efficient. Boring into the ice with metal or diamond drill bits was deemed too slow—it would take an entire summer to drill just one hole. In any case, it's impractical to use mechanical drills to make a hole half a meter wide. Blasting with explosives is also not an option. The only solution is hot water. But, as Duling put it, "water in one of the coldest environments on Earth is a horrible combination."

A veteran driller who had been with the team for many years, Duling provided essential context. First, the hot water pouring out of the nozzle has to carry an astonishing amount of energy (a combination of its temperature and pressure) to melt the ice. "The nozzle puts out approximately sixty-seven hundred horsepower—that's about as powerful as the biggest railroad engine on Earth," he said—and as a former conductor on the Burlington Northern, he knew what he was talking about. But melting the ice is not enough; the drill has to keep plunging down. The IceCube setup uses gravity as an ingenious aid. The drill is weighted with more than half a ton of steel, so it sinks straight down like a plumb bob as the ice melts beneath it, deviating less than a meter from the vertical by the time it has reached the bottom. It takes twenty-four hours to descend 2.45 kilometers and ten hours to come back up, using more than 5,000 gallons of fuel, every ounce of which has to be flown in on U.S. military cargo planes.

Second, working with water at the pole means that time is always critical, as I was about to find out. Duling was getting ready to start the drilling when the group of DVs turned up, escorted by Al Elcheikh, whose normally cool demeanor was about to be tested.

"This is it?" said one incredulous DV, looking at what was just a fat hose going down a hole—after all, the whole project was costing upward of $270 million. Elcheikh, understated and restrained, nodded and proceeded to list the bare-bones details of the drill—the depth, the time it took to get all the way down, and so on.

Someone else asked whether this was a twenty-four-hour operation.

"Yes," said Elcheikh. "Three shifts, ten people per shift."

The DVs milled around, asked some more perfunctory questions, commented on the fact that the project was on time and on budget, and then decided to leave. "Thank you, gentlemen," said a DV. "Good luck."

As soon as they left, Duling called out, "I want this in the water as fast as we can get it there."

There was some tension in the cabin. They had lost precious time. Duling wanted the drill and a submersible pump to be lowered all the way into the hole. Any delay could be disastrous. The water, which was being pumped into the hole at 200 gallons a minute, was coming from two tanks, which held a total of about 15,000 gallons. It was imperative that the meltwater from the ice being melted be pumped back into the tanks. If the holding tanks were depleted, the entire drilling operation would have to stop until the tanks were filled up again, very slowly, from something called a rodwell—the standard way of making water at the South Pole. A rodwell is created by drilling into the ice down below the firn line. Hot water is pumped into this hole, and it starts melting the ice. Some of this water is extracted for the needs of the station, even as more hot water is pushed back in to create more fresh water. But such a well can provide only the seed water to get the drilling started. (IceCube has its own rodwell and doesn't draw from the South Pole station's much bigger one.) Now that drilling was in progress, the pump had to go into the hole and start recirculating the meltwater if the operation was to continue.

"We are on a 200-gallon-a-minute deficit right now," said Duling. "It's a balancing act."

The drill was still above the firn line. The water being poured in was disappearing into the snow. The drill had to get past the firn line, into the ice. As it sank, the men got ready to lower the pump behind it.

"Minutes count," said Duling. Eventually the drill crossed the firn line, and he breathed easier. "Running out of water is just the death," he said. "That won't be the case today." Now any meltwater would stay in the hole and could be pumped out.

"We won today, Al," said Duling, leaning out of the TOS cabin and shouting over the din.

The pump followed the drill. Duling carefully watched a computer screen that showed just how much water was above the pump. Nothing so far. Then suddenly the pump hit water. "It's climbing, finally," he said of the column of water above the pump. In minutes, the reading shot up from 0.21 meter to about 1.97. "I want two more meters," he shouted to the crew lowering the pump. Soon the water height above the pump reached 5 meters. "All right, I'm going to fire up the pump. They need the water." The holding tanks had only about 5,000 gallons left, but now that the pump was cranking away at 3,000 rpm, the danger had passed.

"I'm going home," said Duling. The members of the next shift had arrived, and they would continue the drilling. Beer and Kessler whiskey awaited Duling's crew in the IceCube Lounge.

But the race was still on. Once a hole is drilled, all the equipment—the hoses, cables, pumps, half a ton of steel pipes, everything—has to be moved to the site of the next hole, which is usually about a few hundred yards away. This means disconnecting the hoses and temporarily stopping the hot water circulating through them. The men spooled the main hose back onto a giant reel on skis. It was still full of hot water and weighed a formidable 45 tons. A Caterpillar tractor came by to help drag it to the new site. Meanwhile, everyone else huffed and puffed to get the other, smaller hoses and cables across the snow, either by using smaller tractors and Ski-doos or just dragging the equipment themselves.

The overwhelming impression was one of great urgency. The men scampered from place to place, stumbling in the snow, pulling at one section of hose and then another and being careful not to get a foot caught under one of them. The exercise looked more chaotic than it actually was; in about an hour, they had managed to get everything lined up near the new tower, ready to be connected to the hot-water plant. Any longer and the big hose would have started freezing, and there isn't a building at the pole big enough where the spool can be thawed.

Watching them, it was easy to forget that they were working in full ECW gear. Physicist Gary Hill, also of UW Madison, pointed out how much of a chore it was each day just to get dressed for work. "I've got the fleece pants, Carhartts, thick socks, heavy boots, which are difficult to walk around in, couple of layers of thermal underwear, a jacket that goes over, couple of layers of gloves," he said. "I don't like the wind on my face, so I put on a hood, a polar fleece, a tube to breathe through, a helmet for protection. And then I have my goggles."

The clothes are a burden when walking, let alone staggering around in ankle-deep snow lugging heavy cables and pipes. Hill, who was highly respected by the drill team for doing his share of the hard manual labor, said after one such hose-dragging session: "My god, that's the most exhausted I have ever been. You have the heavy machinery that moves the hoses around, but still you've got to man-handle them a bit. Just picking them up and dragging them, you get out of breath. And the altitude adds to it. With the low air, it's hard to concentrate and think straight sometimes. After work, you feel lethargic and you want to veg out. Here you can't watch TV, so you put in some video you've watched a hundred times and watch it until you fall asleep." This from a man who regularly plays Aussie-rules football in Wisconsin and hits the gym every day, even at the South Pole.

Once the TOS is cleared of all the drilling equipment, it's time for a different team to start lowering one of the strings: sixty digital optical modules connected to a 2.5-kilometer-long cable. This team, too, has to work swiftly, because the water in the hole begins freezing within hours of the drilling. Computer simulations show that in less than a day and a half the hole has narrowed too much to allow the DOMs, which are 13 inches in diameter, to pass through. If the team successfully deploys the string, eventually the water turns to ice and the DOMs are forever frozen in place.

The man in charge of lowering strings was Thomas Ham, a mild-mannered oil worker who had worked in Louisiana, Texas, and Nigeria before he joined IceCube. When I arrived, Ham was going through a safety exercise prior to the deployment, identifying those who would do CPR, pointing out emergency exits in the tower and muster points in case of fire. He turned to me and said, "Your biggest danger is the hole. Don't turn your back on it."

I had no intention of doing so. I stood watching as unobtrusively as I could in a crowded space, as they began lowering the Kevlar-reinforced cable down the hole. Every 17 meters, they connected a DOM to the cable, a slick operation that involved keeping the heavy DOM hoisted on a crane while two of the crew made the electrical and mechanical connections. A team member photographed each DOM as it disappeared down the hole.

Mark Krasberg, a young UW physicist, lifted DOMs off a shelf and lugged them over one by one to the crew near the hole. I noticed that each DOM had a sticker with a name on it. One was named Camembert, the next, Oyster. I inquired about these weird names. They had been Krasberg's idea. Originally, the DOMs all had serial numbers, but Krasberg soon realized that nobody remembered the numbers, especially if they had to recollect a particular DOM that showed up faulty during testing. But it was hard to forget something named Salmonella or Influenza. Of course, those naming the DOMs got carried away, too, with tongue twisters that only a scientist could love, like Aeronausiphobia (the fear of being sick on a plane) or Andromimetophilia (attraction to women who imitate men).

A boombox kept the crew company. It was an odd sight: big men in greased red parkas, thick yellow gloves, and heavy boots swaying to music as they worked on their own piece of the deployment puzzle. Suddenly someone started reminiscing about the music Ham had inflicted on them last year: Elton John, Lyle Lovett, and Randy Newman. No one had minded that too much, until the soundtrack from Disney's *The Little Mermaid* filled the tower operating structure's small cabin. The memory of this left them nearly speechless. This was a crowd more at home with the song playing at the moment:

plaintive strains of Willie Nelson's "Whiskey River" — "*Whiskey River, don't run dry / You're all I got, take care of me.*" It seemed a strangely appropriate song to be playing at the bottom of the world.

The day the New Year arrived, BK told the thirty or so people who had assembled around the South Pole geographical marker, "[Anyone] can watch the ball drop in Times Square on television, but we are the only ones who actually get to see the South Pole marker moved." Despite the numbing cold, everyone stood by to witness the unique celebration. A staffer drilled two new holes in the ice about 9 meters from the old geographical marker and the wooden plaque honoring Scott and Amundsen. Then, to loud cheers, the marker and the plaque were yanked and moved to their new site.

Of course the geographical South Pole hadn't moved, but the ice sheet above it had. The East Antarctic Ice Sheet moves about 9 meters each year, as it slides relentlessly down to the coast. Each year, the folks at the U.S. Amundsen-Scott South Pole Station put in a specially machined marker at 90° south. The marker moves with the ice, so in a year's time it, too, is 9 meters away from the South Pole. Hence the annual ceremony to relocate the marker made by those who wintered the previous season at the South Pole. This year, the marker was a gleaming brass disk the size of a very large hockey puck, with fifty-four notches along the circumference of the disk, one for each person who stayed at the pole during the winter of 2007.

This was the fifty-first move of the marker since it was first set up in 1956, when Rear Admiral George Dufek became the first man to set foot on the pole following Scott's departure in January 1912. Dufek was the commander of a task force in Operation Deep Freeze, the U.S. Navy effort that established bases in Antarctica. Dufek and six others flew in on a Douglas DC-3 passenger plane specially equipped with skis and jet-assisted takeoff (JATO) bottles filled with rocket fuel. The plane, called *Que Sera Sera*, landed at the pole on October 31, 1956. It was a brave, even foolhardy thing to do, for the plane was operating at its very limits. And landing was the easier task — they still had to take off, and no plane had ever done so on ice at such high altitude and in such cold. The temperature was –58°F. The men stayed on the ice for less than an hour before clambering back into the plane. "Conrad (Gus) Shinn revved up the engines, which had remained running throughout, for takeoff—and nothing happened!" wrote Geoffrey Lee Martin in *Hellbent for the Pole*. The skis were frozen to the icy surface, and "Shinn had to fire, successfully, all 15 JATO bottles before the aircraft shuddered free and staggered barely at flying speed into the air." Within weeks of Dufek's pioneering flight, the U.S. Navy established the first buildings of the South Pole station. Dufek had brought an end to the heroic era of Antarctic exploration and ushered in a new one. "He called the stations they established at the South Pole and elsewhere in the Antarctic 'beachheads' which were built and made ready for the 'occupying forces'—trained scientists whose only opponent would be the unknown," Martin wrote.

One of the biggest unknowns confronting IceCube is what's happening near the bottom of the ice sheet, where IceCube's strings lie. "In the worst case, the bottom is frozen stuck to the bedrock and the top is moving at nine meters per year, so there is a shear," explained Ryan Bay, a UC Berkeley physicist. Bay was at the South Pole to attach inclinometers to a few strings; over the next few years, the inclinometers will map the movement of the ice at the bottom. If it turns out that the ice at the bottom is shearing strongly, it could seriously damage IceCube's strings by skewing the lower regions relative to the upper, possibly even snapping the cables in the decades to come, thus reducing IceCube's longevity.

But why lower the strings all the way to the bottom, if shearing ice is a concern? The answer dates back to the first neutrino telescope built at the South Pole: AMANDA, the Antarctic Muon and Neutrino Detector Array. The basic principles, common to both detectors, were formed in the 1990s by Francis Halzen, Robert Morse, and their colleagues. The key was to adapt the technology developed by the team at Lake Baikal to the ice. The concept was simple: Drill holes in the ice using pressurized hot water, lower strings of sensors into them, and let the holes refreeze. Even as others were planning elaborate neutrino detectors in lakes and seas in the Northern

Hemisphere, the South Pole seemed to offer a better alternative, as the thick ice sheet offers a stable working platform. There is no need for ships or underwater vehicles and no worries about the strings being buffeted by undersea currents. Plus there are no concerns about slime growing on detectors or light pollution from radioactivity in seawater or bioluminescent creatures. What could go wrong? Well, the ice, for one thing. What they learned is how little we actually know about ice and how it behaves.

The Antarctic ice was thought to be free of air pockets or bubbles below a depth of 400 meters. But when the first AMANDA strings were lowered to a kilometer, the physicists were in for a shock. The ice was full of bubbles, which caused the Cherenkov light produced by neutrino interactions to scatter after traveling very short distances. This in turn made it hard to discern the original direction of the neutrinos. So the team drilled deeper and lowered another set of AMANDA strings to a depth of 2 kilometers. Down there, the pressure from the ice above turns the bubbles into crystals of air hydrate clathrate, which have nearly the same refractive index as ice, so there is far less scattering. AMANDA proved that an ice-based detector works: It has spotted large numbers of atmospheric neutrinos. However, AMANDA failed to glimpse any astrophysical neutrinos before its reign as a standalone experiment ended in 2005. The stage was set for IceCube.

With its eighty strings carrying a total of four thousand eight hundred light detectors to monitor a cubic kilometer of ice, the completed telescope will be nearly one hundred times the volume of AMANDA. But IceCube sought clearer ice and longer strings. Would it be better to drill even deeper into the Antarctic ice sheet? To find out, the researchers lowered three of AMANDA's strings to a depth of 2,300 meters. They discovered that the ice was indeed clearer. Meanwhile, an ice core drilled at the Lake Vostok station, higher up on the Antarctic Plateau, had revealed clearer ice at a depth of 2,450 metres. So IceCube scientists decided to drill deeper than they had for AMANDA, certain that they would find clearer ice at those depths at the South Pole. Their hunch proved right.

Once the IceCube team drilled the first hole, a team led by Buford

Price of UC Berkeley carried out a detailed analysis of the ice, using a tool called a dust logger. As this device is lowered down a hole in the ice, it fires a laser and analyzes the reflections. The dust logger showed that there are fine layers of dust within the ice, deposited over tens of thousands of years, some due to ancient volcanic eruptions. But below a depth of about 2,100 meters, the device found very little dust. The ice at 2,450 meters is more than sixty thousand years old—and crystal clear.

Still, drilling that deep was extraordinarily ambitious, and the proposal to do so met with intense skepticism. "Everybody told us that IceCube was a cute idea that wouldn't work," said Halzen. Happily for him, the skeptics are being proved wrong.

Within minutes' walk of the drill camp is a mesmerizing reminder of just how well IceCube is working. It's a display inside the Ice-Cube laboratory, a building that would look at home in a colony on Mars. It's a metallic-blue rectilinear structure with rounded edges and egg-shaped portholes of darkened glass. The entire building is on stilts and is connected by walkways to two giant white cylindrical structures, which are the conduits for the cables that snake out from the strings in the ice. A neutrino telescope starts functioning the moment the first string becomes operational, and additional strings serve to increase the volume of ice being monitored.

The display is a graphic on a computer screen showing, in near real-time, the muons that are streaming through the ice. The ice is an inky black background, against which the IceCube strings are a line of dots, each dot a DOM frozen in the ice. The sky is a checkered pattern of blue lines. Every few seconds a muon comes streaking down from the sky, sometimes at an angle to the strings, sometimes straight down. The DOMs register the muon as it passes by, and they light up in one of five colors: red, orange, yellow, green, or blue. The closer a muon is to a DOM, the redder it looks; the farthest muons are blue. The greater the energy of the muon, the more the DOM flares up. I knew it was just a visual representation of what was happening kilometers beneath the surface, but it was hypnotic.

Low-intensity muons caused the DOMs to light up a little, making them look like pearls. High-intensity muons caused nearby DOMs to become so big and bulbous that they overlapped and stuck together, like lumps of gummy candy. The display rotated slowly to provide a 3-D view, a psychedelic soiree of subatomic particles.

By the end of the austral summer of 2008–09, IceCube was at three-quarters the intended size, already bigger than any neutrino detector ever built. And even though it has yet to see any astrophysical neutrinos, IceCube is picking up atmospheric neutrinos, and when complete it expects to see fifty thousand of these each year, compared to about twelve hundred for AMANDA.

It's this ability to detect enormous numbers of atmospheric neutrinos that is laying the foundation for studying the effects of quantum gravity. As we saw in chapter 3, neutrinos come in three flavors—electron, tau, and muon—and as they travel through space they can change from one type to another. This is a quantum-mechanical effect, something that stunned physicists when they first discovered it. Neutrinos, it turns out, are always in a superposition of states, each with a different mass. These mass states have slightly different quantum-mechanical frequencies, and when a neutrino is flying through space these different frequencies interfere, causing the neutrino to oscillate between one form and another. Smaller neutrino detectors, such as Super-Kamiokande in Japan, have seen this effect in atmospheric neutrinos.

AMANDA and IceCube, however, are seeing atmospheric neutrinos of much higher energies than Super-K, and at such energies the atmospheric neutrinos run out of room to oscillate; the distance between where the neutrino is generated and where it is detected is simply too short. AMANDA, for instance, does not see any atmospheric neutrino oscillations. But with IceCube's increased numbers and sensitivity, Halzen and his Wisconsin colleague John Kelley are hoping to see an entirely different kind of oscillation, due to another quantum effect: the interaction between the structure of spacetime and neutrinos. This would clearly indicate the violation of Lorentz symmetry. Because neutrinos are a mixture of more than one quantum state, each with its own frequency, each frequency will interact differently with the structure of spacetime. If Lorentz symmetry is violated, each frequency could have a different maximum velocity. This, too, will cause the neutrinos to oscillate from one type into another. At least that's what *should* happen if spacetime at the Planck scale is not smooth. With fifty thousand high-energy atmospheric neutrinos per year, IceCube will have the numbers to answer such questions.

Fermilab's Dan Hooper has identified another possible test for IceCube, and it has to do with cosmic rays. Normally, cosmic rays are composed of charged particles, which are deflected by intervening magnetic fields, so it's difficult to work out their source. But if astronomers can find cosmic rays coming directly from some source, it will most likely be made of neutral particles, mainly neutrons. (There is speculation of just such a source in the Cygnus region of the Milky Way, but it remains controversial.) When neutrons decay, they produce electron neutrinos. These neutrinos will oscillate as they travel through space, and some of them will have turned into tau and muon neutrinos by the time they reach Earth. Hooper's calculations for a potential source in the Cygnus region show that the neutrinos are expected in ratios of 56:24:20—for electron, muon, and tau neutrinos, respectively. But if spacetime is foamy, then these neutrino types should interact with the fabric of spacetime and "decohere," leading to an equal mix of all three types. So, if IceCube were to see an equal split of neutrino types from the Cygnus region instead of the 56:24:20 ratio expected of smooth spacetime, it would be hard to explain without invoking quantum gravity. But Hooper is aware that such a measurement is a long way off, given that we have yet to see even a single neutrino from such distant sources. Also, the source has to be distant enough for quantum decoherence to become a factor—and theories are of no help at this time in predicting just how far away a source of electron neutrinos needs to be to allow us to observe this effect. It will take some luck for IceCube to stumble on the right source at the right distance. "Stars have to align a little bit, and sometimes they do," said Hooper, somewhat wistfully.

Despite such difficulties, phenomenologists continue to think of ways in which IceCube could detect quantum-gravitational effects. And there's yet another way: the detection of neutrinos from powerful sources halfway across the universe, such as gamma-ray bursts (GRBs). These neutrinos would also interact with the structure of spacetime. GRBs are thought to spit out lower-energy neutrinos first and then higher-energy neutrinos, so the lower-energy kinds should arrive first on Earth. But if they travel through spacetime that is frothy or discrete at the Planck scale, neutrinos with energies of about, say, 1,000 teraelectron volts (TeV) will interact differently with spacetime than 1-TeV neutrinos will, and the higher-energy particles will accelerate relative to the lower-energy ones. The difference in arrival time of the two kinds of neutrinos could be as much as 1,000 seconds. "Suppose you see the high-energy neutrinos before the low-energy neutrinos—that's a beautiful signal right there," said Halzen.

IceCube will coordinate its efforts with space-based gamma-ray telescopes like those aboard NASA's SWIFT and FERMI satellites. These telescopes detect gamma-ray photons and can provide an accurate measure of the distance to the GRBs. If IceCube collects neutrinos from several GRBs that are at varying distances from Earth and finds that the amount of acceleration or deceleration of neutrinos depends on the distance they have traveled, it will be a clear indication that quantum-gravity effects are playing a part. Standard astrophysics cannot explain such behavior.

When Super-Kamiokande and the Sudbury Neutrino Observatory demonstrated that neutrinos have mass (see chapter 3), it was a challenge to the standard model of particle physics, which says that neutrinos have no mass. It was thus clear that there was physics beyond the standard model. Halzen hopes IceCube will shed more light on exactly what lies beyond. When neutrinos smash into the ice and spit out muons, the collision is presumed to follow the rules of the physics of weak interactions, as set out in the standard model. But what if there are subtle things about the collision that are not following the script? Halzen thinks the sheer number of neutrinos that IceCube will see, along with the increased energy of these neutrinos, will give them the statistical clout to ask such questions. "This is a hunt for new physics," he said.

IceCube has turned the icy wastes of Antarctica into a neutrino physicist's field of dreams. "Everybody thinks something will be seen at a cubic-kilometer scale," said Halzen. "I'm not responsible for what is emitting neutrinos in the universe. We developed the technique, we built it, and we'll see what comes."

IceCube is venturing into the unknown. It's a discovery instrument, not unlike the proposed Square Kilometre Array or the Large Hadron Collider near Geneva, Switzerland (my next stop). These behemoths will serve physics for decades, peering through the mists of time to let us glimpse our very beginnings.

Edward Wilson, who accompanied Scott to the South Pole and died with him on the way back to McMurdo, might as well have been speaking about the IceCube scientists when he penned these words during his brief sojourn in Antarctica:

> And this was the thought that the Silence wrought
> As it scorched and froze us through
> For the secrets hidden are all forbidden
> Till God means man to know,
> We might be the men God meant should know . . .

Back at the South Pole station, after a long day out on the ice, I wandered into the hydroponic greenhouse. Days of dry air had left me longing for moisture. As soon as I entered the greenhouse—a small room with row upon row of small plants (clover, sprouts, basil, cress, mustard greens, arugula, parsley)—I was assailed by the smell of fresh, wet leaves. It wasn't emanating from any plant in particular—it was just a dense, moist smell.

Nearly 99 percent of Antarctica is devoid of vegetation. There are no smells—besides diesel, food, and people's bodies in the populated parts. You're not really conscious of this until you leave. Thomas Ham turned poetic when it came to describing the feeling after months of living on the ice. "When the airplane lands in Christchurch, it's normally evening," he said. "They land and open the door, and you can smell the moisture in the air. You can smell the grass and the eucalyptus. When you get off that plane, it just about overwhelms you. You can smell the Earth."

I was ready for that smell. My time at the South Pole—and, indeed, in Antarctica—had come to an end. Days later, sitting in the plane, ready to fly to Christchurch, I realized that I had not touched the Antarctic snow with my bare hands. I stepped off the plane, removed my gloves, and picked up a fistful. I was reminded of something NASA glaciologist Robert Bindschadler had said earlier. He spoke of the "absolute stillness" he felt in Antarctica, the absence of anything familiar. "You are there in the middle of nature, and nature is expressed in very simple, impressive, and intensive terms. You can't ignore the temperature, because it is cold. When the wind blows, you can't ignore the wind, because it dominates your awareness. And the surface that you are standing on and walking across and living on, that's all there is. It's omnipresent and it's overwhelming. There is no other distraction. I feel I can't avoid the issue of our place in nature and our uniqueness and smallness."

9 · THE HEART OF
THE MATTER

THE SHORT BUS RIDE from Geneva took us through fields of corn and sunflowers. It was still early morning; the sky was overcast. The sunflowers looked like little people, their yellow heads bowed, waiting for their first glimpse of the sun. It felt quiet, this pastoral Swiss setting. Yet some 300 feet beneath this land, a giant machine would soon rumble to life, generating energies not seen since moments after the big bang.

The bus made its final stop outside CERN, the European Organization for Nuclear Research, which straddles the border of France and Switzerland. A row of flags of European nations fluttered in the light breeze. In the far distance loomed the sheer face of Mont Salève, a 4,500-foot limestone massif that towers over Geneva.

I was on my way to see the 7,000-ton ATLAS, a particle detector being built for the Large Hadron Collider (LHC), the world's largest particle accelerator. The LHC smashes protons together at energies never before achieved on Earth, creating particles that could have existed only in the infant universe, when the cosmos was hot enough for them to pop in and out of existence. And then detectors like ATLAS sift through the debris of particles, hoping to find answers to a multitude of questions. How do elementary particles get their mass? What is dark matter? Are there extra dimensions? Is there a whole new set of particles besides the ones we already know about? Will these particles lead us toward a theory of quantum gravity?

The watchword in these experiments is precision. This is perhaps what makes Switzerland a perfect country to host CERN. The legendary Swiss obsession with accuracy is apparent the moment you land in Geneva. Inside the airport, instead of posters of movie stars or sports icons, you are greeted by mammoth wall-to-wall posters of the world's most sophisticated chronometers. One manufacturer has cashed in on the region's fame for physics and cosmology by calling one of its watches the Big Bang. And just before you exit the airport, a giant billboard reminds you that you are only five minutes away from CERN, the world's largest physics laboratory. It's a stunning aerial view, with the suburbs of Geneva and the farms of Switzerland and neighboring France in the foreground and snow-covered Alpine peaks in the background. Overlaid on this landscape is the arc of a gigantic circle, depicting a portion of the underground tunnel that houses the Large Hadron Collider. If you were to draw the markings of a clock on this giant imaginary circle, Geneva's airport would barely span the notches from noon to two. There's no experiment as big as the LHC anywhere else in the world.

Particle accelerators are the crown jewels of physics. In them we confirm our most far-reaching theories. Astronomers and cosmologists might discover dark matter or dark energy, or even find indirect clues to the nature of these mysterious components of the universe, but until physicists create the predicted particles in colliders and verify their properties, skeptics will not cede ground. Similarly, theoreticians can invoke the presence of extra dimensions to elegantly explain the cosmos, but until the signs of these hidden parts of the universe manifest themselves in detectors, questions will persist. Particle colliders and detectors are where the loose ends are tied and theories verified beyond doubt. "It takes machines like the LHC to journey to the edge of our understanding because, quite simply, the easy stuff has all been done," Brian Cox, a physicist working on the ATLAS detector, wrote in an article in the *Guardian*.

The edge of our understanding in physics has reached a point where it has become necessary to probe energies that existed a billionth of a second after the big bang. To do this, the LHC accelerates two beams of protons in opposite directions to within a whisker of the speed of light. At this speed, a proton will go around the LHC's 27-kilometer-long tunnel 11,245 times each second. Ground tides caused by the new moon and the full moon can cause the tunnel to shrink by 1 millimeter, and since the energy of the beams depends on the number of times they go around the tunnel, even this minuscule effect can change the energy of the protons. When these particles smash together, all that energy is concentrated into a region of space thousands of times smaller than the width of a human hair, creating the energy densities needed to spawn new particles that could prove or disprove theories.

CERN, with its rich history of particle physics, is uniquely poised to leap into the unknown. My walk to the ATLAS offices took me past a veritable graveyard of early particle detectors. These instruments, some weighing up to 26 tons, had played their part in the study of elementary particles at CERN during the 1970s and 1980s. After decades of service, they had become technological headstones, paying homage to a long tradition of probing the very essence of matter. I wound my way through CERN's outdoor cantina, probably past some of the brightest minds in physics today (nine thousand scientists from eighty-five countries do research here), and came to a confluence of small roads named after several giants of the past — Marie Curie, Pierre Auger, Wolfgang Paul. In front of me was Building 40, which houses most of the scientists working on ATLAS.

In this supercharged environment, where every turn reminds you of particle physics, a statue on one side of Building 40 seems both anomalous and appropriate: a 6-foot-high cast-metal figure of the Hindu god Shiva. It stands at the end of a gravel-strewn path. In this particular form, as Nataraja, the king of dance, the multi-limbed, long-haired Shiva is dancing inside a giant ring of fire. His right leg is trampling the dwarf of ignorance; the left is raised, knee bent, in a posture of delicate balance. Nataraja dances to end one cosmic cycle and beget the next, a metaphor for the cyclical nature of the universe in Hindu cosmology. The historian Ananda K. Coomaraswamy once wrote of the dance:

> Shiva rises from his rapture and, dancing, sends through inert matter pulsing waves of awakening sound. Suddenly, matter also dances, appearing as a brilliance around him. Dancing, Shiva sustains the world's diverse phenomena, its creation and existence. And, in the fullness of time, still dancing, he destroys all forms — everything disintegrates, apparently into nothingness, and is given new rest. Then, out of the thin vapor, matter and life are created again. Shiva's dance scatters the darkness of illusion, burns the thread of causality.

The fact that creation and destruction coexist is only too apparent at CERN. Machines built here since the 1970s have been smashing particles together, annihilating them into pure energy, and then analyzing the new particles created in the cauldron. During the past decade, however, CERN has witnessed another kind of death and rebirth — that of the particle colliders themselves.

On the day of my visit, the place was humming with the efforts of physicists and engineers striving to put the finishing touches on the LHC and its various detectors. The LHC is a once-in-a-generation machine. But before it could be built, an older accelerator, the Large Electron-Positron (LEP) collider running in the same tunnel, had to be torn down. (Such machines are built underground not for safety reasons but because it is cheaper to dig gigantic tunnels than to pay for the real estate on the surface — especially in prime locations like Geneva.) By 1999, the LEP was nearing the end of its life, but it refused to die a quiet death. The machine had come tantalizingly close to finding an elusive particle called the Higgs boson, the one undetected member of the so-called zoo of particles described by the standard model of particle physics. The Higgs is thought to give elementary particles their mass. Finding it is essential for a complete understanding of the material world. So, teased by the promise of the LEP, CERN officials gave it more time to find the Higgs while the construction of the LHC proceeded apace.

Creation and destruction. Civil engineers could not afford to wait until the LEP had been dismantled to build the cavern to house ATLAS. ATLAS was too big. So for a few months a carefully choreographed dance of extreme engineering ensued. While the LEP was still running underground, gigantic machines began boring their way down from the surface, toward the LEP. One of the most delicate operations—a strange phrase when you consider the scale of the engineering involved—was the excavation of two 60-meter-deep shafts and a cavern beneath the shafts. This was where ATLAS would live. The cavern had to be 35 meters high, as tall as a ten-storey building, but it was directly in the path of the LEP's electron and positron beams, or "beamline." Engineers started excavating the shafts, using mechanized rock-breakers rather than explosives, so as not to disturb the sensitive LEP beneath.

Normally when you build a cavern, you excavate all the way down to the base, lay the floor, build the walls and then the ceiling. But because the LEP was still running, the engineers had no choice but to cast the vaulted ceiling first. It was no ordinary ceiling. At 10,000 tons of concrete and steel, this massive 2-meter-thick structure had to be built and then held in place by thirty-eight cables bolted into special structures halfway up the shafts. Eventually the LEP was shut down (without having found the Higgs), and the engineers started excavating the rest of the cavern with the ceiling hanging over them. Nearly 300,000 tons of rock were hewn out, the floor was laid, and the walls were erected. At last, the ceiling had the necessary support beneath it. When workers removed the cables holding the ceiling, the civil engineering had been so precise that the ceiling barely moved as it settled on the walls.

None of this drama was apparent when I entered the ATLAS cavern for the first time. The almost fully assembled detector was so huge that it stole the show. It was only when ATLAS spokesperson

Peter Jenni told me that the cavern actually moved that its size registered. Here was a cavern that pushed upward against nearly 60 meters of solid rock above and moved—as a bubble of air would in water—despite the 7,000-ton detector inside it. The amount of rock excavated to house ATLAS was far greater than the weight of ATLAS, so the "bubble" in the rock was considerably less dense than the surrounding rock, causing it to rise. "This was predicted by the civil engineers," said Jenni. Indeed, the cavern was moving about 0.2 millimeter upward a year. Every piece of precision equipment to be installed in the cavern had to be designed to accommodate this movement. Most important, the cavern floor was a whopping 5-meter-thick layer of steel-reinforced concrete, ensuring that as the cavern moved up, the entire floor moved uniformly, without warping.

I stared at the behemoth that was ATLAS—all steel girders, aluminum supports, cables, muon chambers, silicon trackers, liquid helium pipes, and superconducting magnets—and realized I would have to come back again to make any sense of it. All I could take in at the time was that the LHC was going to smash protons against protons at four points along the ring and ATLAS was at one of those points—like a gigantic cocoon around a microscopic region of space that would soon witness energies never before created on Earth.

To understand just how important the LHC is to particle physicists and cosmologists, one needs to understand the problems with the standard model of particle physics. The model, which divides the material world into fermions (the building blocks of matter) and bosons (the carriers of fundamental forces), has been amazingly successful at explaining subatomic particles and their interactions. But over the past few decades, there have been strong hints that it is incomplete.

One glaring omission in the standard model is that it does not incorporate gravity and hence does not predict a particle for this fourth fundamental force. But that's just one of its problems. Take the Higgs boson, a particle required by the model. This particle is crucial to explaining the origin of mass—not the universe's mass but the mass of elementary particles, such as electrons and quarks. Most of the mass of atoms in the universe consists of protons and neutrons (electrons contribute less than one part in a thousand to the mass of atoms). Protons and neutrons, in turn, get their mass from their constituents, quarks. The quarks are held together by particles called gluons, the carriers of the strong nuclear force. It is the "binding energy" of this force that gives protons and neutrons their mass, and it accounts for 90 percent of the mass of normal matter. Except for one big problem: The standard model forbids quarks and gluons (and, indeed, all other elementary particles, including the electron) from having any mass if it weren't for the Higgs boson.

The explanation goes something like this: The Higgs boson is associated with a Higgs field, just as the electron is associated with an electromagnetic field, but with one major difference. An electromagnetic field is a vector field, which means that at every point in space it has an associated direction. The Higgs field, however, is scalar—it has a value but no direction, not unlike temperature in a room. Think of the Higgs field as a thick, gooey background against which all other particles have to move. Every other fundamental particle interacts with the Higgs field in its own specific manner, and the strength of this interaction gives each particle its mass. By all accounts, the Higgs should have been found by now. That hasn't happened.

Then there is something called the hierarchy problem. The Higgs boson gets some of its mass from its interactions with every other particle. Add up these contributions, and the Higgs mass turns out to be nearly 10^{17} times greater than what experiments say it should be. To rein in the mass of the Higgs, the various contributions—some positive, some negative—have to cancel out exactly, and for that to happen some parameters of the standard model have to be fine-tuned to an astounding precision of one part in 10^{34}. "We believe that fine-tuning is an undesirable property," said Dan Tovey of the University of Sheffield, U.K., a member of the ATLAS team. "Obviously we don't want the mass of particles to be what we observe just because of some remarkable coincidence."

Physicists hate coincidences. They would like all aspects of nature to emerge naturally as solutions to their equations. And that's where an extension to the standard model called supersymmetry comes in. An English theorist, John March-Russell of Oxford University, who was at CERN when I visited, recounted the feverish excitement over supersymmetry that gripped particle physicists in the 1990s, when the LEP was running at full tilt.

The first phase of the machine, called LEP1, had identified the strengths of the electromagnetic force and the weak nuclear force at the machine's operating energy (that is, the total energy of the particles colliding head-on). In the case of LEP1, the electrons and positrons smashed into each other with an energy of about 91 giga-electron volts, or GeV. (By contrast, the LHC has an operating energy of 14 teraelectron volts, or TeV.) Theorists think that the fundamental forces were unified as one force in the first instants of the universe. So if you took the strengths of three of the four forces—all except gravity—found by the LEP and other experiments and extrapolated them to the higher energies that existed moments after the big bang, you would expect them to unify—that is, become indistinguishable. While this turned out to be nearly so in the standard model, there were discrepancies that left theorists uneasy.

A new idea called supersymmetry came to the rescue. It posits that every particle has an as-yet-undiscovered superpartner. Specifically, all bosons have fermionic partners, and all fermions have bosonic partners. March-Russell recalled that some physicists objected to supersymmetry, arguing that doubling the number of particles was ridiculous. But it wasn't absurd to many other physicists, including March-Russell, who calls supersymmetry "a very beautiful and powerful idea." For one thing, the doubling of particles eliminated the hierarchy problem. Using supersymmetry, when physicists extrapolated the strengths of the forces as determined by experiments to the higher energies of the early universe, the forces unified very accurately. "'Very accurately' means to one part in two thousand," said March-Russell. "[And] there is no hierarchy problem. You can reinterpret this as a prediction [of supersymmetry]." The finding impressed many. It was incredible that this beautiful idea could predict the unification of the electromagnetic force and the weak and strong nuclear forces so precisely. It would be a tremendous red herring if it weren't true, said March-Russell.

And that's not all. By the mid-1980s, dark matter had become a serious issue for astronomers. It turned out that the lightest supersymmetric particle—the neutralino—was a likely candidate. As noted in chapter 2, whereas nearly all the predicted supersymmetric-partner particles are heavy and unstable and hence would have decayed instantly in the infant universe, the neutralino is stable. It would interact weakly with matter, just like dark matter. Here was a perfect candidate for 90 percent of the matter in the universe. Even more impressive were supersymmetric calculations showing that once the universe had cooled enough for all particle annihilations to have stopped, the density of leftover neutralinos—known as the relic density—would be just right to give the universe the dark matter that is observed today.

Surely this beautiful theory would be proved by observations of supersymmetric particles at the LEP. Encouraged by the performance of LEP1, CERN decided to boost the energy of the collider to 209 GeV. While the LEP was being upgraded, theoretical physicists were getting quietly confident. Until then, all their predictions using the standard model had been verified by experimentalists. Now not just the Higgs but even supersymmetry was within reach. However, they remained cautious, at least publicly. "The honest truth is that we tended to say, 'It has a good chance of seeing the Higgs,' but we didn't tend to talk about supersymmetry," said March-Russell. "That was more speculative, and it wasn't appropriate." It was just as well that the theoreticians kept their excitement to themselves. LEP2 found neither the Higgs nor any sign of superpartners.

What LEP2 had been looking for were particles predicted by the simplest version of supersymmetry, called the minimal supersymmetric standard model (MSSM). As with the standard model, MSSM has its share of tunable parameters. Given the energies at which LEP2 had searched for supersymmetric-partner particles, and given that it had failed to find them (or, indeed, the Higgs), most of the values of these tunable parameters were ruled out. If you think of the parameters as an n-dimensional matrix—called the parameter space—with the value of each parameter varying along one axis, then most of the matrix had been explored and nothing had been found. Only a tiny corner of the matrix remained. "Supersymmetry wasn't in the main region where we expected it to be," said March-Russell. "It is still this beautiful idea, but we have to explain why we are in this strange region [of parameter space]."

Theorists, after decades of successfully—some would say arrogantly—telling experimentalists what they would find in their particle accelerators, were finally stumped. "In a way, it is a good situation," said March-Russell. "Theorists felt they had the upper hand. Basically, since about 1975 they had been leading the experimentalists." By the late 1990s, that had changed. Now theorists are uncertain as to just what the LHC is going to find.

The LHC and its detectors are ready to explore the energy regions that LEP2 could not. Will they find the neutralino? If they do, it will be the first serious identification of a dark-matter candidate. But it will also open up our world to supersymmetry. Interestingly, the standard model predicts a single Higgs particle, while supersymmetry predicts multiple Higgs particles. What kind of Higgs will the LHC find? As loyal as they are to the standard model, theorists are still seduced by supersymmetry. "If I had to bet the house, I would bet on supersymmetry," said March-Russell.

Fabiola Gianotti, a slightly built, dark-haired Italian physicist who took over from Peter Jenni as the spokesperson for ATLAS on March 1, 2009, is equally optimistic. "If supersymmetry exists at the TeV scale, it will be discovered at the LHC," she said, as we sat in her office at CERN, the wall behind her desk dominated by a poster of a tropical island with palm trees silhouetted against an orange-red sunset. "Actually, it will be one of the easiest new things to discover, even easier at the beginning than the Higgs boson."

Physics wasn't Gianotti's first love. "I came to physics from very far away," she told me. "When I was a young girl, I loved art and music. I had been studying piano quite seriously at a conservatory and had taken courses in high school targeted towards literature, languages like ancient Greek and Latin, philosophy, and history of art. I loved these subjects, but I was also a very curious little girl. I was fascinated by the big questions. Why are things the way they are? This possibility of answering fundamental questions has always attracted me—my mind, my spirit, everything."

She stumbled upon physics soon afterward. "I discovered that physics is really interested in the most fundamental questions," she said.

More than philosophy?

"Even more," she said, speaking slowly to emphasize each syllable. "Because experimental physics is based on facts. It is answering fundamental questions—not just giving an answer to your question by inventing something, but proving it. This is very, very nice."

This was no theorist talking. Here was someone who got down-and-dirty with instruments. These concepts—supersymmetry, dark matter, the Higgs, extra dimensions—were not mere equations to her but ideas that left traces in her instruments, whether in the form of streaking jets of particles or in some anomalous measurement of momentum or energy.

Gianotti recalled the first time she went down into the ATLAS cavern. "I said to myself, 'My god, how are we going to fill it? It is too big.' Now when I go down, I say, 'My god, how are we going to fit everything? It is so small.'"

But everything did fit, and today, with the LHC switched on, protons are colliding in the center of ATLAS. Somewhere in those collisions are what particle physicists call signatures—maybe of supersymmetry and dark matter, or of the Higgs.

Physicists talk of such signatures with so much certainty that it's easy to construe them as permanent, tangible evidence. But in reality these signatures are fleeting. In the LHC, for instance, when protons collide with protons, the high energies, for the briefest moment, produce conditions that spawn particles and their antiparticles, as decreed by quantum mechanics. These particle/antiparticle pairs can either annihilate each other or individually decay into other particles, each less massive than its predecessor. The higher the energy of the initial collision, the greater the mass of the particles that could emerge from it. But there's no way of actually seeing the first particles produced; their lifetimes can be as short as 10^{-24} seconds. So the physicists have to study the stream of secondary particles created as the primary heavyweights decay, concentrating on the familiar end products from the world we know, such as quarks, electrons, muons, and photons. Then they work backward, trying to figure out which primary particles might have generated the secondary ones.

A Higgs boson, for instance, theoretically can decay into four electrons or four muons or two electrons and two muons, among many other possibilities. A Higgs can also decay into a pair of bottom quarks. These are the simplest scenarios. The search for the Higgs gets much more complicated in the world of supersymmetry, where most models predict not one but five Higgs particles, three neutral and two charged, each with its own decay chain.

A supersymmetric universe could reveal itself in other ways. The simplest would be missing energy. If a massive supersymmetric particle is created, it will instantly decay into lighter and lighter particles until the decay chain ends with the production of the lightest supersymmetric particle, the neutralino. The neutralino is by definition stable and weakly interacting—making it a choice candidate for dark matter. But because it's weakly interacting, a neutralino will slip through an instrument like ATLAS undetected. Or almost undetected. The laws of physics say that the total energy and momentum is conserved during these collisions. So if you monitor every single particle in a given decay chain, the neutralino will make its presence felt as missing energy and momentum.

Missing energy and momentum could also signal the presence of extra dimensions. A major problem with the standard model is that it cannot explain why gravity is so much weaker than the other fundamental forces (for example, it's nearly 10^{34} times weaker than the electromagnetic force). String theorists have suggested that all the forces are equally strong—it's just that gravity operates in extra dimensions, whereas the other forces are locked into the three dimensions familiar to us. So gravity seems weaker, most of it leaking away into the extra dimensions. It all sounds like science fiction until you hear physicists like Gianotti explain how extra dimensions could turn up at the LHC. The collider could produce gravitons — particles that are thought to carry the gravitational force. These gravitons will escape into extra dimensions, but the LHC has a chance of spotting them — or, as with neutralinos, inferring their absence in the form of the energy or momentum missing at the end of a long and complicated chain of decaying particles.

Before beginning its search for new physics, however, ATLAS will have to confirm what other detectors have found. The Tevatron collider and its detectors at Fermilab, near Chicago, have established to considerable precision the masses of the W boson and the top quark (which is one of the six types of quarks, and the heaviest), while the LEP did the same for the Z boson. ATLAS will have to find them, and not just that — the physicists will have to use these particles to fine-tune their new detectors, making sure they are working as designed. The top quark, discovered at Fermilab in 1995, is one of the particles physicists will use to calibrate ATLAS. "The signal of today becomes a calibration tool of tomorrow," said Gianotti.

The LHC and ATLAS could uncover some deep truths about the universe. Gianotti confessed to "feelings of excitement and the awareness of being close to something very important and great for humankind." She quoted the thirteenth-century Italian poet Dante Alighieri: "Fatti non foste a viver come bruti ma per seguir virtute et conoscenza" (We were created not to live as animals but to pursue virtue and knowledge). "As human beings, the pursuit of fundamental research and knowledge is a need for us, which separates us from animals or vegetables. It is like the need for art," said Gianotti.

For a man who studied aerospace mechanics, Francois Butin seems extraordinarily at ease underground. In 1992, fresh out of college and required to put in some sort of service for the French military, Butin was offered the choice between working for an oil company in war-torn Angola or working at CERN. Understandably, Butin chose CERN for a sixteen-month stint, and he never left. By 1994 he was working for ATLAS.

Inside a building blandly called the SX1 Surface Hall—a wooden structure designed to blend in with pastoral Switzerland but whose real purpose is to provide cover for the activities beneath it—Butin pointed to a satellite map of the region, with Lake Geneva, the Jura Mountains, and the Geneva airport clearly visible. On it were overlaid the various underground tunnels and caverns that house the LHC and its detectors. The 27-kilometer LHC ring crosses the French-Swiss border four times. ATLAS is located in Meyrin, Switzerland. Diametrically opposite, at the base of the Jura Mountains, is a competing detector, built inside French territory, called CMS (for Compact Muon Solenoid). There are two other detectors, called ALICE and LHCb. These four detectors, all of which are in caverns that lie 300 feet below the surface, envelop the four points at which the counterrotating beams of the LHC cross and collide. If you think of the beams as trains, then ATLAS, CMS, ALICE, and LHCb are underground train stations through which the trains pass and, sadly, collide. ATLAS analyzes the aftermath of such collisions.

Also in SX1 is a large cutaway poster of ATLAS, which looks like a cylinder lying on its side. Even on its side, it's as high as an eight-storey building and almost twice as long. Underground, the LHC's proton beams streak along the axis of the cylinder, crossing and colliding at the very center. The particles created by these collisions—a menagerie that includes, among others, quarks, electrons, muons, and photons—are studied in exquisite detail by layers of electronics that surround the center. At the very center is the inner detector, which measures the momentum and charge of every particle created. Just outside it are the calorimeters, which measure the energy of every particle except muons and neutrinos. Being electrically neutral and extremely elusive, neutrinos stream out of ATLAS without being detected. Muons are charged, however, and banks of muon chambers positioned outside the calorimeters track them. Determining the momentum and location of each muon, along with the energy and momentum of every other particle, is critical to figuring out what exactly happens when protons collide at the heart of ATLAS.

But why is ATLAS so big? It all comes down to magnets. They are the key to tracking particles accurately. The properties of charged particles can be calculated by seeing how they bend in a magnetic field. The stronger the magnetic field, the greater the deflection of the particles produced in a collision and the greater the precision with which the particles can be studied. One method for increasing the strength of magnetic fields is to wrap the current-carrying coils of a magnet, or solenoid, around an iron core. But an iron core has one disadvantage—the muons can hit iron molecules and get scattered. Any muon chamber measuring the tracks of these particles has to accommodate this scattering, which reduces its precision.

To avoid this problem, ATLAS was designed with two magnets. A central solenoid (without an iron core) generates a small but strong magnetic field at the heart of the detector. The inner detector, which keeps track of all particles as soon as they emerge from the collisions of protons, is fitted inside the magnet. The calorimeters, which measure the energies of particles, are just outside the magnet. The superconducting core of the central solenoid is as thin as possible, to minimize scattering.

But the real challenge is to track the muons that make it past the inner detector, central solenoid, and calorimeters. To do this, the designers went for an air-core toroidal magnet—essentially, massive coils of superconducting wires filled with nothing but air. There is no metal inside the coils. "The idea was to create a huge magnetic field volume with virtually no material inside," said Butin, as we stood inside SX1, above one of the 60-meter-deep shafts. An air core means that there will be no scattering of particles, making the measurements of muon trajectories very precise. But to get magnetic fields of the desired strengths without an iron core, this magnet has to be extremely large. The engineers finally ended up with eight gigantic coils of superconducting cables; if you stood one on its end, it would be taller than a seven-storey building.

Preventing muons from scattering was not the only thing on physicists' wish list. They also wanted to avoid any dead spaces — that is, regions in a detector that are not actively tracking particles. By definition, any support structures holding up the detectors are dead spaces. If a muon goes only through the support structures, avoiding any contact with the detectors, there's no way for physicists to reconstruct its path. The solution is to minimize the number of support structures — no mean feat when you have to prop up thousands of tons of electronics. "These guys wanted to design a dream detector for physicists," said Butin. "What they did not know was that they were designing a nightmare for engineers."

Peter Jenni had emphasized the same point. "The importance of engineering cannot be stressed enough for actually realizing the dreams of physicists, who forget that you cannot just design an instrument — you have to support it somehow," he said. "And supports are a nuisance, because they take up space." But just how much of a nuisance? The search for supersymmetry, for instance, hinges on a detector's ability to precisely pin down the energy and momentum of all the particles that pass through it. A smidgen of missing energy and momentum could signal the existence of neutralinos (and thus dark matter) — or gravitons, which would hint at extra dimensions. But what if the missing energy or momentum is the result of some particles passing through support structures rather than the detectors? "That's dangerous," said Jenni. Thus you have to be careful about what you conclude.

Butin shepherded me toward the elevator that would take us down to the ATLAS cavern. When CERN invited bids for its construction, an Italian company, noting that the elevator had only two stops, bid to finish building it in eight weeks. "They had not read that it was ninety-two meters deep," said Butin. Six months later, they were fired.

Eventually the elevator was completed, and I was now coming to the first stop, 81 meters below the surface. We exited into a cavern called the Universal Services Access (USA)-15 Hall, which housed all the equipment and electronics that would have to be accessible when the proton beams were colliding in the nearby ATLAS cavern.

The radiation from the collisions would be so intense that the main cavern would be out of bounds. Inside the hall, the first thing I noticed was the noise. The clanging of machinery, the hum of electronics, the deep rumble of helium compressors, the sound of drills and saws, the grind of metal against metal, all melded into an indistinguishable roar.

The hall housed an entire electrical substation—6 megawatts of power to be fed to ATLAS. The superconducting air-core toroidal magnet alone requires 20,500 amperes of current. That is so much electricity that if the magnet has to be shut down, the orderly process of draining the current from its coils takes about three hours.

We walked past the substation into a room full of electronics. This was ATLAS's nerve center. The computers on the racks are responsible for separating the electronic wheat from the chaff. To emphasize just how critical these computers are, Butin took a moment to explain exactly what happens when the proton beams collide at the center of ATLAS. The beams are not a continuous stream of particles; rather, each beam is made up of bunches of protons. Going around the tunnel will be 2,808 such bunches in each direction. When they cross paths, at four specific points, only about twenty of the 200 billion protons that make up the two bunches will collide; the rest will continue their journey around the accelerator unimpeded. Not much to keep track of, one might think—except when you consider that the beams are moving at near the speed of light, and thus 40 million bunches cross every second. Their collisions produce 100 billion particles per second.

The ATLAS computers sift through this extraordinary barrage of data, looking for interesting collisions that could contain signatures of the Higgs, a neutralino, or extra dimensions. The whole thing is like a conveyor belt moving at nearly the speed of light. The computers are barely 500 feet from where the protons collide. Even as the computers are analyzing the events from one collision, data from other collisions are on their way. In the meantime, muons are streaming past the inner detector, and the muon chambers are working feverishly to track some and ignore others. The system cannot detect all the muons—there are just too many of them—so it depends on feedback from the computers to decide which ones are more important. The phrase "clockwork precision" takes on a different meaning here: Seconds, milliseconds, microseconds are too long. We are in the realm of nanoseconds. Spacetime has been sliced into the thinnest of slivers.

Leaving the electronics behind, we came to red steel doors; beyond them was the main cavern. When ATLAS is running, these doors are locked and the cavern is out-of-bounds. Massively thick concrete walls separate ATLAS from the rest of the facility to shield those working nearby from radiation. We entered the cavern at the base, on to one of the thirteen storeys of metal walkways that ring the inside of the cavern, reminiscent of walkways in high-security prisons in Hollywood movies. The steel clanged underfoot. An empty cavern must have been an awesome sight, but right now it was upstaged by this mass of metal and electronics: ATLAS. The cavern had been usurped by its occupier.

Most of the 3-D jigsaw that is ATLAS had been assembled underground, its pieces lowered one by one down the two enormous shafts overhead. The inner detector, the central solenoid, and the calorimeters were all in place. So was the air-core barrel toroidal magnet. The superconducting coils were enclosed in metal tubes—cryostats, designed to keep the coils cold—which fanned out around the detector's central core.

Almost everything in the cavern had been designed with disaster in mind. The eighty or so people assembling the detector all wore harnesses; all had been trained to work high above the ground, since the risk of falling was great. I was allowed to walk only along the metal walkways, holding on to the railings, and even so, my knees would occasionally soften as I glimpsed the dizzying drop beneath me.

There was also the danger of gas leaks. "You have ninety tons of liquid argon in this cavern," said Butin. "Enough to fill the whole cavern if it leaks." While argon is inert and nontoxic, in high enough concentrations it can reduce the amount of oxygen available and cause people to suffocate. In case of such a leak, the cavern has giant fans to suck it out; their strong blades are made of stainless steel so that they won't buckle as they're expelling the heavy gas.

Then there is the risk of fire. Despite the extreme care that the engineers have taken to minimize the use of any flammable material, ATLAS still has 1,800 miles of cables. While they are halogen-free, designed not to generate toxic and corrosive smoke while burning, they can still burn, and most of them are concentrated inside the detector's inner core. "Firemen cannot reach them easily," said Butin. "If we have a fire we cannot control, we can fill this cavern with foam." He pointed to enormous blowers high up on the cavern walls, which can spew out foam—literally, by mixing "bad-quality soap" with water—and fill the gigantic cavern in ten minutes, depriving the fire of oxygen.

What if people are in the cavern at the time? "There is enough air in the foam to breathe," said Butin. Anyone inside the cavern would have to stay still and breathe normally. "It is very uncomfortable, I can tell you, but there is enough air to survive," he said. "If you want to be comfortable, you do this." He pulled his shirt over his head. "It filters out the foam, and you get just the air."

ATLAS is audacious. How can an experiment so large still maintain the precision required to measure particle paths down to distances of tens of microns, barely the width of a human hair? Take the giant muon chambers. Toward one end of the cavern, workers were assembling a circular end-cap, a muon detector that would slot into one end of the cylindrical ATLAS. The end-cap was being constructed by piecing together giant trapezoidal sections. A boom crane stretched up from the bottom of the cavern to the top of the end-cap; at the very end of the boom was a cage with a lone engineer inside. I could barely see him at this distance. Within each trapezoidal section were individual channels that would detect muons. Each was a pressurized gas-filled tube with a wire running through it. A muon passing through would ionize the gas, creating a current in the wire. Nearly a million such channels made up the muon-detecting system in ATLAS, and the position of each wire inside each channel had to be monitored to an accuracy of about 25 microns.

How do you do that, when the aluminum support structures are constantly expanding and contracting with subtle changes in temperature, thus shifting the channels' positions ever so slightly? The megawatts of power used by ATLAS also generate heat, and while most of it is siphoned away by the cooling systems, some of it heats the air. An assembly of cameras, temperature sensors, and lasers work in concert to keep tabs on the channels. You need to know precisely where they are if you want to track particles down to a precision of tens of microns. Unreasonable numbers—thousands of tons, millions of channels, and widths of just a few microns—coexist inside the cavern in mind-numbing harmony.

"You are not afraid of heights?" Butin asked, snapping me out of my micron musings. We were climbing to the top of ATLAS. I denied any fear, kept my misgivings to myself, and walked along the steel scaffolding, stepping over masses of cables and wires, to reach the very top of ATLAS. Above us was the 10,000-ton vaulted ceiling that once had hung from cables. Everywhere I looked were gigantic metal beams that were supporting the entire detector. The toroidal magnet's magnetic field—which, needless to say, wasn't switched on—is so intense that a 10-ton beam feels the force of an additional 10 tons bearing down on it. Nowhere else on Earth has a magnetic field so strong occupied a volume so large. It takes all these materials' strength to keep them from collapsing.

The noise of the construction, the clanging of metal, and the whine of the vacuum pumps and helium compressors were nearly unbearable. When the LHC is running, there's a new sound: the roar of the proton beams themselves entering the cavern—and the thunderclap of protons smashing into each other. In 2004, engineers tested a prototype proton beam slamming into a target. The test was necessary to develop the technology to stop the real beam in a hurry in case it went awry. This is an accelerator physicist's nightmare. A lost beam—essentially, particles that cannot be kept on the straight and narrow as they go around the tunnel—is dangerous. When running at full throttle, each of the LHC's counterrotating proton beams has an energy of about 360 megajoules, equivalent to the energy of a 400-ton high-speed train traveling at 150 kilometers per hour. Such an errant beam, which is capable of melting a 500-kilogram block of copper, has to be stopped instantly, or it can wreak havoc on the delicate instruments, literally liquefying anything directly in its path. It can take weeks if not months or years to repair the damage. In the 2004 test, physicists redirected a 450 GeV beam into an auxiliary tunnel and dumped it into a target. Microphones picked up the gunshot-like bang. If the two LHC beams, each at 7 TeV, have to be dumped, they will smash into their target with the sound of 150 kilograms of TNT.

We climbed down from ATLAS and slipped into a bypass tunnel that led into the main LHC tunnel. Thick concrete walls separated us from ATLAS, and now it was blissfully quiet. Butin pointed to the beam pipe. The LHC's protons are confined to this pipe as they race around the ring. The vacuum inside the pipe has to be extreme or the protons can smash into molecules of air, wasting energy and even spewing out particles in regions of the beam pipe not equipped to handle them. So, besides sucking the air out of the beam pipe with powerful pumps, the engineers coated the insides with a substance called non-evaporable getter; most of the remnant air molecules stick to this coating. The result is an intense vacuum—you would have to go 600 miles up into space to find similar emptiness. Butin spoke almost reverently of the vacuum, his voice hushed. "Here we have vacuum that we cannot even measure," he said.

"What do you mean?" I asked.

"It is so low that when you bring a probe inside, it can never be clean enough not to pollute what you have in there," he said. "We don't know how to measure the vacuum."

If the vacuum in the beam pipe is unearthly, the temperature of the magnets that control the beam is more so. Unless an alien civilization has built something colder, the LHC is the coldest ring in the universe.

The temperature of outer space is about 2.7 K. That's –270.3°C, or just 2.7 degrees above absolute zero. Inside the 27-kilometer tunnel at CERN is a piece of equipment that is colder than outer space, making Pluto seem like the tropics.

More than any other aspect of the LHC, the temperature of its main magnets embodies the challenge of building a new accelerator in the tunnel that once housed the LEP. The LHC is designed to work at 14 TeV, about seventy times the energy of LEP2. The higher the energy of the collision, the greater the chance of creating heavier and heavier particles that otherwise could never be seen. But reaching collision energies of 14 TeV inside the old LEP tunnel hasn't been easy. That's because an accelerator's energy is related to the radius of its tunnel and the strengths of the magnets that keep the charged particles confined to the tunnel. The LEP tunnel had been designed for energies of a few hundred GeV. One way to build the more powerful LHC would have been to dig an entirely new and much longer tunnel. But that had already been attempted elsewhere, and the harsh lessons were there for all to see.

The Superconducting Super Collider (SSC) was designed by American physicists to achieve energies of about 40 TeV, and the ring was to be 87 kilometers in circumference. The site chosen was near Waxahachie, in northeast Texas, about 50 kilometers south of Dallas. The ambitious plan was approved in 1987 by President Ronald Reagan, who used a popular sports metaphor from American football to rally the physicists (a tribe that really needs no such encouragement). "Throw deep," he said. Unfortunately, it was a throw that sailed past the end zone. By 1993, the cost estimates for building the SSC had ballooned from $4.4 billion to more than $12 billion. The U.S. Congress canned the project, leaving behind a 22.5-kilometer stretch of completed tunnel that now lies derelict. Physicists around the world were shocked and dispirited. The letdown inspired novelist Herman Wouk to write *A Hole in Texas*, which follows the travails of a fictional physicist who had hoped to find the Higgs boson at the SSC and whose life is derailed by the decision to kill the project.

Determined to avoid the SSC's fate, CERN decided to use its existing LEP tunnel and cut civil engineering costs. But doing so meant that the magnets had to work tremendously hard to steer the LHC's beams, which would be an order of magnitude more energetic than what the tunnel had been designed for. Steering a beam is not unlike driving a Formula One racing car. The faster the car, the harder it is for the driver to tackle tight bends. F1 drivers train hard to build up their arm and (especially) neck muscles to handle the turns. Magnets are the muscles of the collider world. Of the more than 9,000 superconducting magnets inside the LHC tunnel, 1,232 need special mention. These are dipole magnets, and they are the machine's neck muscles. Each weighs 35 tons, and the entire lot has to be cooled down to 1.9 K, the temperature of superfluid liquid helium. Helium gas liquefies when cooled down to 4.2 K at atmospheric pressure. But take the temperature down further and liquid helium becomes a superfluid, with some unusual properties. One property is its extraordinary thermal conductivity. "It is ten thousand times better than oxygen-free copper, the best copper you can get," said Lyn Evans, the LHC project leader, in his singsong Welsh accent, as we walked through the LHC tunnel.

Standing next to one of the gleaming dipole magnets inside the tunnel, Evans expounded on the virtues of superfluid helium. Its high thermal conductivity means that it can whisk heat away from the magnets, but superfluid helium is also a nuisance. It has zero viscosity and can slip through the most microscopic of cracks, so the plumbing that carries it has to be infallible. "The quality of the many, many thousands of welds has to be as good as those in a nuclear plant, at least," said Evans. The plumbing will keep the superfluid helium flowing between the magnets and the refrigeration systems, maintaining the superconducting coils all around the tunnel at 1.9 K.

The extreme temperature makes the LHC the world's most challenging machine. Repairing it is far from trivial. "It's like being in outer space," said Evans. "If anything goes wrong, you can't just go in there and touch it." First, the magnets have to be warmed back up to room temperature, which takes about five weeks. After repairs, the LHC's 40,000 tons of magnets need to be cooled back down to 1.9 K, which takes another five weeks and requires nearly 10,000 tons of liquid nitrogen and 130 tons of superfluid helium. Given such constraints, "the quality control has been draconian on this machine," said Evans.

Unfortunately, the quality control wasn't good enough. On September 10, 2008, CERN fired up the LHC, and nearly a billion people around the world tuned in to witness the maiden voyage of protons around the tunnel. But two days later, a power transformer failed, so the LHC was shut down for a week. Within days of restarting, an electrical fault blew a hole in the plumbing and some of the liquid helium instantly vaporized, leading to a rapid buildup of very high pressure, which caused severe mechanical damage to nearby equipment, including some of the magnets. Evans's prophetic words came true. It would take more than a year before the LHC team could repair the damage and begin colliding protons.

Despite its teething troubles, the LHC has avoided the fate of the Superconducting Super Collider. The engineers have done their bit. Now it's time for the science. As of this writing, the last of the damaged dipole magnets have been repaired, lowered into the tunnel, and reinstalled. Meanwhile, another team is three-quarters of the way through cleaning the beam pipes, removing all the metal debris and soot from the electrical fault—a far more involved procedure than it sounds. It requires brushing the beam pipe mechanically and vacuuming it and repeating the procedure ten times before checking the insides of the pipe with an endoscope. Other teams are developing new tests to prevent the kind of disaster that stalled the LHC in the first place.

Of all the people who were waiting for the LHC to start up, one man could rightfully lay claim to being the most eager: Peter Higgs. In April 2008, Higgs, a seventy-eight-year-old British physicist, came to CERN for the first time in decades, just before the caverns and tunnels were closed to the public, to see for himself the machine that could find his eponymous particle. It was in the 1960s that Higgs had invented the Higgs mechanism, which endows elementary particles with mass, and after decades of hype and hoopla the particle is now close to being found. There's little doubt in his mind that the Higgs particle (or particles) will be discovered soon.

The day after he toured the LHC, the usually reclusive Higgs faced a gaggle of reporters in Geneva. He is famously modest, refusing to refer to the Higgs particle by name, preferring to call it the "boson named after me," and remains careful to point out that he wasn't the only physicist to come up with the idea. Two physicists in Belgium, Robert Brout and François Englert, discovered it almost simultaneously; it is now officially called the Brout-Englert-Higgs mechanism. But Higgs did make one unique contribution: he predicted a particle that could be found by experiments. "The only reason I suggest that maybe I deserve to have the particle named after me is because I drew attention to it," said Higgs.

The Higgs boson gained wider notoriety when Nobel laureate Leon Lederman wrote a 1993 book about it called *The God Particle*. Lederman had wanted to call the book *The Goddamn Particle*, but his publisher's commercial instincts won out and the book ended up with the more grandiose title. An atheist, Higgs finds it extremely embarrassing that the boson has such a nickname. Still, what if the LHC finds it? "I shall open a bottle of something," said Higgs. "Whisky or champagne?" quipped a media wag. "Champagne," said Higgs. "Drinking a bottle of whisky takes a little more time." More poignantly, Higgs was asked if he thought the LHC would find the boson before he turned eighty. "I hope so," he said. "[Or] I'll just have to ask my GP to keep me alive a bit longer."

While the discovery of the Higgs boson will surely land him a Nobel Prize, Higgs said he would be far more excited if the LHC found signs of supersymmetry. Why? Because the route to a unified theory of all four fundamental forces, and hence to a theory of quantum gravity, seems to go through supersymmetry.

Steven Weinberg knows something about unifying forces. He received the Nobel Prize in 1979, along with Sheldon Glashow and Abdus Salam, for his work on the unification of the electromagnetic and the weak forces. We met at his office at the University of Texas in Austin, in the fall of 2007. It was a typical theoretician's office, with books piled high everywhere and a blackboard chalked with inscrutable equations. Out of the corner of my eye (not wanting to stare), I spied a photograph of Weinberg standing next to royalty, most likely taken during the Nobel Prize ceremony. Weinberg was categorical about the implications of discoveries at the LHC. "If the LHC simply discovers a single Higgs boson and nothing else, well, they will open bottles of champagne, of course, but it will be a disaster, because it will mean that we have no clue about how to go forward," he said. "If they discover signs of supersymmetry, that will be enormously exciting."

One undisputed beneficiary of such a discovery will be string theory, which requires supersymmetry for its equations to work. But since nature can be supersymmetric without being stringy, the discovery of supersymmetry alone will not be enough to convince skeptics of string theory's validity. Nonetheless, any boost for string theory in its current form means a nod to the idea that our universe is one of many universes (the multiverse), and consequently a nod to the anthropic principle. Weinberg will be unhappy if the answers to some of our biggest questions turn out to be anthropic. "I take negatively to it, in the sense that I hope we will be able to calculate everything from first principles, without the use of anthropic arguments," he said. "I would be very happy to throw over anthropic reasoning, if I had any ideas on how to do it."

Still, Weinberg argues for keeping an open mind, both about the anthropic principle and the multiverse—but always with his tongue firmly in cheek. In 2005, on his way to Trinity College at Cambridge University from Austin to deliver the opening talk at a symposium titled "Expectations of a Final Theory," Weinberg spotted an issue of *Astronomy* magazine in which English astronomer Martin Rees was quoted as saying that he was confident enough about the existence of a multiverse to bet his dog's life on it and Stanford cosmologist Andrei Linde added that he was ready to bet his own life on it. Weinberg retorted, "I have just enough confidence about the multiverse to bet the lives of both Andrei Linde *and* Martin Rees's dog."

Given that so many lives are at stake, experiments like the LHC are being avidly monitored by theorists. While discoveries at the LHC will be key, many theorists will also be following the outcome of experiments being conducted in the most extreme environment of all—the cold, dark emptiness of space.

10 · WHISPERS
FROM OTHER UNIVERSES

ON FEBRUARY 1, 2007, the European Space Agency (ESA) unveiled the Planck satellite, the latest in a small but select group of pathbreaking space probes designed to map the cosmic microwave background. The CMB is, literally, the universe's first light. A fossil radiation more than 13 billion years old, it contains imprints of processes that occurred moments after the birth of the universe. Its discovery in the 1960s was considered decisive evidence in favor of the big-bang theory—no other model of the universe can produce such radiation.

Scientists studying the CMB are the glittering stars of cosmology, having garnered a clutch of Nobel Prizes, so it was perhaps fitting that Planck was introduced in Cannes, France, a city famous for its glitzy film festival. That day, the proverbial red carpet had been laid out, at least in spirit, for Berkeley cosmologist George Smoot, who was still basking in his newly minted status as a Nobel laureate for his work on COBE, the COsmic Background Explorer satellite. He spoke at length about the virtues of Planck and how it would take us into a golden age of precision cosmology, in which the parameters that describe the universe, its structure and evolution, would be measured with greater and greater accuracy. Then someone from the audience asked, "What is your confidence in the big-bang theory? Did the big bang ever happen? Are you really sure?"

Smoot was taken aback. "I'm surprised that this question has been asked in France—in the United States I wouldn't be quite as surprised," he said. "I have a very high degree of confidence in the big-bang theory. We don't have as much evidence [for it] as we have for the theory of evolution, but there are people in the United States who question the theory of evolution, too. Ever since the Nobel Prize announcement, I get five times as many e-mails [saying], 'Well, it's great you got the prize, but it was a mistake. The big bang is wrong, the radiation isn't what it is.'"

Smoot was justifiably thrown by the question, for he had just reminded his audience that satellites preceding Planck—COBE and WMAP, the Wilkinson Microwave Anisotropy Probe—had already confirmed, with astounding accuracy, some of the predicted characteristics of the CMB.

Until about 370,000 years after the big bang, the photons in the universe were trapped, unable to go anywhere because they kept bumping into free electrons. Then, at the epoch of recombination, the universe had cooled enough so that the electrons and nuclei combined to form neutral atoms, mainly of hydrogen. The universe became transparent to light. Photons could finally travel through space unhindered. Those photons, which have themselves cooled to near absolute zero because of the expansion of the universe, give us a snapshot of the universe at recombination.

But the CMB is a window into an even earlier era, when the quantum fluctuations of our infant universe were stretched to cosmological scales by inflation in the first fractions of a second after the big bang. By the time of recombination, these fluctuations had become tiny variations in the density of matter. When the CMB photons were set free, they scattered off matter for the very last time before beginning their journey across the cosmos. Those that scattered off denser regions were colder than those that came off less dense regions. That's because they had to expend energy trying to escape the gravity of the denser regions. The difference in temperature induced by these variations is as little as one part in a hundred thousand. COBE saw these variations and created a Monet-like painting of the microwave sky. And WMAP—which has thirty-three times better angular resolution than COBE—confirmed it with far greater precision, constructing intricate pointillistic images.

Planck can map the temperature variations, or anisotropies, in the CMB in still more detail. In fact, the instruments on the satellite are as good as they get for measuring anisotropies, so in a sense Planck will have the last word on them. Also hidden in these anisotropies is the imprint of the primordial sound waves that began ringing about 30,000 years after the big bang and stopped soon after recombination (see chapter 6). CMB telescopes, including WMAP, have made a great start at seeing these imprints, but Planck is doing it better.

When the acoustic waves began propagating through the plasma of the early universe, each wave spreading out from an over-dense region of dark matter, their mutual interaction set up standing waves (such waves, for example, are produced when you pluck a guitar string; the string vibrates at its fundamental frequency, or wavelength, and also at all its harmonics). The universe rang out at a fundamental frequency (the biggest wavelength) and its harmonics. The biggest standing wave would have caused parts of the plasma to reach maximum compression and other parts to reach maximum rarefaction at the time of recombination. The harmonics would have caused smaller and smaller regions to reach maximum compression and rarefaction. These patterns of maximum and minimum density were frozen into the fabric of the universe soon after recombination and it is this information that is preserved in the CMB. The photons of the CMB from compressed regions would have been hotter than the photons from the rarefied parts. So if you look at the sky at an angular scale of 1° and less, you should see photons that are hotter or colder than average by about thirty parts per million.

As we saw in chapter 6, the maximum distance an acoustic wave could have traveled before the oscillations stopped serves as a standard ruler in the sky. At recombination, this distance was about 120 kiloparsecs, and in the CMB it should subtend an angle of about a degree—that is, if the universe is flat. Most scientists believe it is, but the accuracy of current measurements is such that there is room for doubt. Is the universe slightly curved? If so, is it negatively or positively curved? Planck can answer these questions by measuring the standard ruler more accurately than ever before.

The probe will also illuminate for the first time the inflationary era of the universe. Inflation is thought to have roiled the very fabric of spacetime, generating gravitational waves, which would have grown in size with the expanding universe. At the time of recombination, these gravitational waves would have simultaneously stretched and squeezed spacetime, twisting and polarizing the photons of the CMB.

Polarization is best understood by thinking of light as an electromagnetic wave that vibrates in a plane perpendicular to the direction in which it's traveling. In that plane, the angle of vibration can be aligned with the vertical or horizontal or anywhere in between and is called the angle of polarization. Usually, photons from a light source are randomly polarized, so the light doesn't have a net angle of polarization. But in some situations light can be preferentially polarized—as when sunlight is reflected off a lake's surface (that's why some anglers wear special sunglasses to filter the glare of polarized light, enabling them to see below the surface for fish spawning on riverbeds). Inflation-generated gravitational waves would have polarized the CMB photons in very specific ways, and this signature could be present in today's sky. Planck, in effect, is equipped with the sunglasses to see (not filter) this pattern of so-called B-mode polarization. But there's one problem. No one knows how strong such a signal might be, so it's a bit like groping in the dark.

If Planck finds B-mode polarization, it will be the first direct evidence of inflation—the "smoking gun," if you will. It will help cosmologists figure out the strength of the scalar field that caused inflation.

This, along with the curvature measurements, will be our strongest clue yet to the nature of inflation. When did inflation start, and how long did it last? Can string theory accommodate it? Answering these questions is crucial to tackling the bigger question of whether our universe is one of many, many universes in a multiverse. Planck could become our best experimental window, so far, to the multiverse.

Back in Cannes, the Q&A session ended without further challenge to the big-bang theory. Everyone was bused to the clean room at Thales Alenia Space, ESA's prime contractor for Planck, to see the satellite itself. We were warned not to touch anything. The telescope, with its 1.5-meter primary mirror, stood assembled, covered in diaphanous pink sheets. It still had to be integrated with the electronics, radiometers, bolometers (instruments that measure incident electromagnetic radiation), and heat shields that stood next to it. The Planck scientists lined up to have their pictures taken in front of their instruments. It was a moving sight; many of them had been working on pieces of the project for more than a decade and were seeing the entire satellite for the first time. "It's a beauty!" exclaimed Jan Tauber, Planck's project scientist.

Planck was launched some two years later, on May 14, 2009, in tandem with Herschel, a far-infrared space telescope. Both were carried aloft on an Ariane 5 rocket that thundered into space from ESA's Spaceport in Kourou, French Guiana. They were released from the launch vehicle about two minutes apart, and both space probes then sped toward prime parking spots about 1.5 million kilometers, or 900,000 miles, from Earth—nearly four times farther away than the moon—a place called L2 (for Lagrangian point 2), one of the "quietest" spots in space.

Only recently have space probes made the long journey to L2. The two most celebrated probes of recent decades—the Hubble Space Telescope and COBE—were both launched into Earth orbit, Hubble about 360 miles up and COBE at 560 miles. Such orbits are easier and cheaper to achieve—and, in the case of the Hubble, accessible for repairs—but they have their limitations. Cosmologists are after increasingly precise measurements, and probes in Earth orbit can suffer from an unstable thermal environment as they pass in and out of our planet's shadow, especially at the lower altitudes. By the 1990s, space scientists knew that their probes had to get a certain distance from Earth if they were going to collect good data. L2 was the perfect destination.

In 1772, the mathematician Joseph-Louis Lagrange identified five points in the vicinity of two massive bodies where the gravitational pull is so weak that a third, much smaller body can sit "motionless" relative to the two larger bodies. This remained of theoretical interest only until the early twentieth century, when astronomers discovered asteroids that were in the same orbit around the sun as Jupiter and either leading or trailing the gas giant. It turned out that the asteroids (the "smaller" bodies) had settled into two of the five Lagrangian points of the sun-Jupiter system (the two "massive" bodies).

In the last decade, scientists realized they could exploit the sun-Earth system by sending probes to one of two other Lagrangian locations to carry out astronomical and cosmological observations. The first point, L1, is the easier to visualize. Draw an imaginary line about 93 million miles long connecting the sun and the Earth. L1 lies between the two, 900,000 miles away from Earth. It is at this point that the gravity of the sun and the Earth nearly cancel each other out. A spacecraft orbiting L1 can maintain its position relative to the two bigger bodies with little effort. It's an optimal position for viewing the sun. In fact, that's exactly where SOHO, the ESA/NASA Solar and Heliospheric Observatory, has been since 1996, sending us astonishing images of our host star. But while L1 is the preferred location for solar probes, it's a terrible place to gather the faint signals of the distant cosmos, because it's overwhelmed by just about everything the sun spews out.

That's where L2 comes in. It lies on the same straight line and at the same distance from Earth, but on the other side. Earth effectively comes between L2 and the sun. Spacecraft that can reach L2 find themselves in a privileged position. First, as with L1, they have to expend relatively little fuel to stay at L2, because of the benign nature of the gravitational field there. Second, Earth blocks most of the solar radiation. Third, the positions of the sun and Earth are fixed relative to the probe, so any polluting radiation from the two bodies can easily be accounted for. Two months after the successful launch, Planck and Herschel each settled into their separate orbits around L2.

The orbit is not the only thing protecting Planck from the sun. As the probe orbits L2, it always has its back to the sun, with its instruments pointed at the darkness beyond. Though outer space is at a temperature of about 2.7 K, the sunny side of the satellite reaches a toasty 300 K (80.6°F). A sophisticated cooling system disperses this heat into the cold vastness of space, bringing the temperature of Planck's instruments down to about 40 K. Active cryogenics further chill the instruments, one of which operates at a frigid 0.1 K—far, far colder than the surrounding darkness. Because Planck is measuring temperature differences across the sky that amount to less than a few millionths of a kelvin, its instruments have to be correspondingly cold, otherwise their thermal and electronic noise would drown out these faint signals.

Planck is not the first CMB probe to orbit L2. WMAP is already there. For the minimum of fifteen months that Planck will be operating—its lifetime is limited by the cryogenics on board—it will be competing with NASA's still-functioning probe. WMAP was designed to last for only two and a half years, but the probe has functioned so well that its mission has been extended. WMAP, unlike Planck, does not use active cryogenics, so nothing besides fuel limits its lifetime. By the time Planck completes its mission, WMAP will have accumulated eight years of data, and that gives it just enough oomph to compete with the better instruments on Planck.

Studying the CMB is an exercise in improving the signal-to-noise ratio. The CMB is the signal; all other radiation in the same frequency range—emissions from instruments, galaxies, and even galactic dust—is noise. Once the noise is well characterized, continued observation helps to increase the signal-to-noise ratio. Even so, the signs of inflation-generated gravitational waves are believed to be beyond WMAP's reach. Planck could become the first experiment to detect them. And when it does, not only will we have direct evidence that inflation—a crucial event in the universe's history—really happened but we will also learn more about the theories that best describe the birth pangs of the universe.

Planck is named after the German physicist and Nobel laureate Max Planck. And for good reason. We owe our understanding of the cosmic microwave background to Planck's work in the early 1900s, when he solved the so-called blackbody radiation problem and gave birth to quantum physics.

A blackbody, in theory, is an object that absorbs all radiation and reflects or transmits none. Such an object, when cold, appears black, hence the name. But a hot blackbody emits thermal radiation, and experiments in the nineteenth century showed that such radiation had a characteristic spectrum. When you plot its intensity at each frequency, then for a given temperature of the blackbody the intensity rises with frequency, peaks, and then falls off. Increase the temperature of the blackbody, and the peak of this graph shifts toward higher frequencies. What the experiments showed is that the shape of this plot and the position of the peak depended only on the blackbody's temperature—but nothing in classical physics could explain the spectrum. Instead, calculations showed that the intensity of the radiation should increase to infinity with increasing frequency, instead of peaking and falling off, as the experiments were showing. Then, in 1901, Planck solved the problem by suggesting that energy is absorbed and emitted in discrete bundles, or quanta, and not continuously, as required by classical physics.

In the late 1940s, George Gamow and his students Ralph Alpher and Robert Herman realized that the universe is a perfect blackbody. As such, any radiation left over from the big bang should have a spectrum that can be predicted by measuring its temperature. Gamow's early calculations suggested that the radiation today should have a temperature of about 5 K. In the 1960s, James Peebles reinvented and refined the theory (see chapter 4). Then the accidental discovery of the CMB using a ground-based radio antenna in 1965 by Penzias and Wilson gave us the first accurate measurement of the temperature of this radiation: 2.7 K. Many balloon- and rocket-based experiments followed, each making measurements at various frequencies. By 1970, the data at low frequencies matched the theory well, but otherwise it was a mess: There appeared to be fifty times more energy at higher frequencies than expected—but the measurements were too imprecise to confirm this. It was then that NASA announced a competition to design a satellite-based experiment to study the CMB. It would prove to be a bumpy ride.

When NASA first approved COBE, in 1982, the space shuttle was to be the satellite's launch vehicle. Then on January 28, 1986, disaster struck. The shuttle *Challenger* broke apart just seventy-three seconds after its launch, and all seven crew members died. NASA grounded the shuttle fleet. But in its enthusiasm for the shuttle program, the agency had already dismantled other rocket launchers capable of carrying satellites such as COBE. Fortunately, the COBE team discovered that enough parts existed to assemble an old workhorse—the Delta rocket—so they scrambled to redesign COBE and launch it on a Delta. The project proceeded with few hiccups, and the satellite was finally flown to Vandenberg Air Force Base in California and readied for launch. Then came another scare. On October 17, 1989, an earthquake struck the San Francisco Bay Area. NASA's John C. Mather, the principal investigator for the Far Infrared Absolute Spectrophotometer (FIRAS), the main instrument on COBE, later remarked that the tremors were felt at Vandenberg, more than 200 miles south of the quake's epicenter, "but COBE was bolted down that day, because two of our engineers had gone off to be married." The bolted-down satellite survived the quake, and on November 18 it was launched into polar orbit, 560 miles up. The launch had august company: Alpher and Herman witnessed the historic event.

In no time at all, COBE allayed all concerns about the anomalous energies at high frequencies. It showed with exquisite precision that the CMB was at 2.725 K in all directions. The intensity of the radiation varied with frequency exactly as predicted. And it took just nine minutes of observation in space to confirm the blackbody spectrum.

In January 1990, at a meeting of the American Astronomical Society, Mather presented the nine-minute spectrum, and the scientists gave him a standing ovation.

The satellite, however, had more work to do. It had to find the variations in temperature across the sky. George Smoot, the PI of an instrument called the Differential Microwave Radiometer (DMR), was leading the effort to find them. By the fall of 1991, COBE had started seeing these anisotropies, but Smoot was worried. "There was just a chance—small, I thought, but big enough not to be ignored—that our map of the wrinkles was an artifact of the torrent of radiation that pours constantly from cosmic activity within the Milky Way," he wrote in *Wrinkles in Time*, his 1993 account of the COBE project. Smoot decided that the only way to rule that out was to study the radiation from the center of the Milky Way, and the best place to do that was the South Pole, because of its altitude, its cold and stable air, and its grandstand view of the galactic center. But Smoot was miserable. He hated the thought of having to endure temperatures perhaps as low as 70°F below zero. It was only after colleagues told him that he could stand anything for a month that Smoot went to the pole and measured the Milky Way's radiation. Finally, he was convinced that COBE's radiation wasn't an artifact.

COBE had indeed given us our first glimpse of the primordial anisotropies—the density variations, or seeds, that led to the large-scale structure we see today. Smoot created quite a fuss at a meeting of the American Physical Society in Washington, D.C., on April 23, 1992, when he announced rather dramatically, "If you're religious, it's like seeing God." Whatever the feeling, the discovery of the anisotropies trumped the discovery of the blackbody spectrum. For the first time, cosmologists could calculate a property called the angular power spectrum. It's a measure of the variation in temperature for each angular scale of the sky. Imagine you are wearing glasses that allow you to see only those CMB photons that are about, say, 30° apart. Measure the amount by which the temperature of the photons deviates from the average temperature of the CMB. Now put on a different set of glasses—say, ones that let you see at angular scales of 10°.

Note by how much the temperature of these photons differs from the average. Do this for all angular scales. Then plot the angular scale on the x-axis and the corresponding temperature deviation from average on the y-axis. What you get is the so-called angular power spectrum of the CMB. COBE measured the power spectrum for angular scales ranging from 7° to 90° — and it agreed with predictions beautifully. This part of the power spectrum arises due to quantum fluctuations that have been stretched by inflation. The satellite's discoveries proved so seminal that Smoot and Mather were awarded the Nobel Prize in 2006.

But the cosmic microwave background was hiding greater treasures. Smaller anisotropies had escaped COBE's gaze, for it had looked at the sky with rather fuzzy eyes, seeing features only at angular scales bigger than 7°. It was time for others to put on better glasses to see smaller features, for therein lay the information about primordial sound waves. In the graph of the angular power spectrum, the temperature fluctuations caused by the biggest standing waves should show up as a bump called the first acoustic peak. The harmonics, too, would have led to hotter and colder photons, but at smaller and smaller angular scales, and they should be visible in the angular power spectrum as second and third acoustic peaks, and so on.

Balloon experiments like Boomerang, which first flew over Antarctica during the austral summer of 1998–99, began hunting for these acoustic peaks. Boomerang could see at angular scales smaller than a degree and brought the microwave sky into sharp focus. On April 27, 2000, the team announced that they had seen the first acoustic peak. For the first time, there was clear evidence of the geometry of spacetime.

Einstein's general theory of relativity says that matter and energy can affect the curvature of spacetime and that photons will follow this curvature. Astrophysicist John Archibald Wheeler put it succinctly: "Matter tells spacetime how to curve, and spacetime tells matter how to move." In a flat spacetime, parallel rays of light will remain parallel as they travel to infinity. In a closed universe (positive curvature), the parallel rays will eventually converge, and in an open universe (negative curvature) they will diverge. The photons of the first acoustic peak can be used to determine the curvature of spacetime. If the universe is flat, the photons should come straight at us and subtend an angle of about one degree in today's sky. If the universe is open, the trajectory of the photons will curve such that they will be closer together than 1°, and if it is closed they will be farther apart. Boomerang and another balloon experiment called MAXIMA (Millimeter Anisotropy eXperiment IMaging Array) measured the hotspots of the first acoustic peak to be about one degree apart. They had found the universe to be flat.

But all was not right. Neither Boomerang nor MAXIMA had seen the predicted value of the second acoustic peak. This was a problem, because the ratio of the heights of the first and second peaks is crucial to calculating how much baryonic matter (protons and neutrons) there is in the universe. Their results were showing more baryonic matter than could be explained by big-bang nucleosynthesis. It would take another year of analysis, and a race with a telescope at the South Pole, to resolve the issue.

The discovery of the first acoustic peak had been a big blow to those working on the Degree Angular Scale Interferometer (DASI, pronounced "daisy") at the South Pole. Physicist John Carlstrom from the University of Chicago was leading the effort to build DASI, which started in 1995. Nils Halverson, a tall, gangly scientist who is now at the University of Colorado in Boulder, was Carlstrom's student at the time. In 1999, he came to the pole to help complete DASI, and during that Antarctic summer he heard the unwelcome news that the two competing balloon experiments had already found the first acoustic peak. But the doubts about the second acoustic peak meant something was amiss. DASI still had a role to play.

By 2001, Carlstrom's team had collected enough data. They assembled at the American Physical Society meeting in April in Washington, D.C., to declare that they had seen the second acoustic peak. Halverson, however, stayed back in Chicago, analyzing and reanalyzing the data until the day before the announcement. Meanwhile, the balloon teams had gotten wind of DASI's discovery, and they too had scrambled to reexamine their data. At the APS meeting, all three teams announced that they had seen the second acoustic peak and that it agreed with predictions—it had just the right height relative to the first acoustic peak. The baryonic content of the universe determined from the ratio of the two peaks was in accord with big-bang nucleosynthesis. The announcement was huge—yet another epochal moment in our understanding of the universe's origins.

Sitting inside the U.S. station at the pole, just a mile away from where DASI stood, Halverson recalled the day the news broke. "It was a very tense time for me as a graduate student. I was the one person on our team who missed the meeting, because I was back in Chicago making sure the data were right. I remember being in my apartment, working over the weekend just nonstop, with very little sleep—it was one of the most active, tense, sleep-deprived periods of my life." A tired and high-strung Halverson woke up the day after the meeting and picked up the *New York Times*. There it was, their story: "Listen Closely: From Tiny Hum Came Big Bang." For Halverson, it was all a little too much. "I had this incredible release, got to tears in my apartment, tears of relief."

In the midst of all this, NASA launched COBE's successor, the Wilkinson Microwave Anisotropy Probe (WMAP) in the summer of 2001. WMAP became the first spacecraft to reach L2, and by 2003 it had started delivering spectacularly detailed maps of the cosmic microwave background, confirming and building upon the work of the balloon-borne instruments and polar telescopes. While no one denies COBE's place in history and WMAP's incredible accuracy, the two satellites tend to color perceptions of microwave background research. "NASA would like you to think that first there was COBE and then there was WMAP," said Halverson. "But there was all this incredible stuff in between. For an experimentalist who has been laboring in the field for a while, there tends to be this institutional memory that forgets about all the history that leads up to these big, mature experiments that come out and do the definitive measurement."

Unfortunately for experimentalists like Halverson, the story seems set to repeat with Planck, even though it will be the European

Space Agency, not NASA, that will likely be collecting the accolades. Smoot, speaking in Cannes, put it poetically: "Planck is the future of looking back to the past."

No one is awaiting Planck's findings more eagerly than Stanford University's Andrei Linde. He is the inventor of chaotic eternal inflation, which is widely accepted as the successor to Alan Guth's "old inflation" theory. Inflation predicts that the photons of the CMB will be polarized—but the levels of polarization depend on the amplitude of the gravitational waves generated during inflation, which in turn depends on the energy density of the vacuum of spacetime during this period. If Planck sees the B-mode polarization, it will lend strong support to Linde's work, bringing inflationary theory from the realm of the metaphysical right down to Earth.

When Guth came up with the theory of inflation in December 1979, Linde was at the Lebedev Physical Institute in Moscow. Guth's theory did not really work—a fact that Guth himself was well aware of. His proposed mechanism for a universe that somehow got stuck in a "supercooled" state of false vacuum and expanded exponentially did not run to completion. But the idea was too good to pass up—it just needed some tweaking. Meanwhile, Russian physicists Alexei Starobinsky, Viatcheslav "Slava" Mukhanov, and Gennady Chibisov had started working on the same problem—though not with the express intention of solving the flatness problem of big-bang theory, as Guth had.

Inspired by his Russian colleagues, Linde developed new inflation in the summer of 1981. His theory would also make the universe expand exponentially, but he could get it to complete. Because of the Soviet Union's isolation, it took until February 1982 for the paper to be published in the West. It first had to be typed out in Russian, the equations inserted by hand, and then signed by five people who could attest that it did not contain any state secrets. This was then sent to multiple agencies for approval, after which Linde had to retype the paper in English, reenter the equations by hand, and then get another series of permissions to put it in an envelope and mail it to a journal. The envelope took yet another month to reach England, for it was opened, sealed, and reopened several times along the way. When it finally reached cosmologists in the West, new inflation took root much as Guth's idea had—that is to say, in no time at all.

But even new inflation had its shortcomings. Theorists immediately figured out that it did not correctly predict the observed large-scale structure of the universe. Linde, meanwhile, had moved on. In 1983, he came up with yet another version, this one called chaotic inflation. The earlier models of inflation—both his and Guth's—had required the universe to be hot before inflation began. Chaotic inflation, however, did not require big-bang-like temperatures as an initial condition. A speck of spacetime—cold or hot—with a mere milligram of matter (or energy) was enough. The key was the distribution of this energy. Inflation would occur if the early universe had a scalar field whose energy density, instead of being in a state of false vacuum, decreased very slowly. As an analogy, imagine a ball rolling down the inside of a shallow bowl. The ball represents the scalar field, and the height of the ball from the base of the bowl denotes the energy density of the field. The field's natural tendency is to move toward a state of lower energy density. Einstein's equations show that if space is permeated by such a scalar field, then space will expand. Crucially, as it expands, the same equations show that space gets viscous, in a manner of speaking, which prevents the ball—i.e., the scalar field—from rolling down as fast as it otherwise would. This means that space is now stuck with a scalar field that is changing extremely slowly. This then causes space to expand exponentially, leading to inflation, which eventually stops when the ball ends up at the bottom of the bowl.

Any ball that rolls down the inside of a bowl will rock back and forth at the bottom, before it comes to a stop. The same is true for a scalar field. Linde found that such an oscillating field would produce elementary particles, these particles would interact with one another, and the universe would get extremely hot. And all this happened when the universe was still only 10^{-35} seconds old. After this, standard big-bang theory takes over. So the high temperature we associate with the big bang occurred at the end of inflation: It is the bang in the big bang.

The new theory got its name from Linde's realization that the first instants of the universe would have been chaotic—in the sense that the scalar field could have had different arbitrarily determined values for its energy density at different points in space, however small the volume. Inflation would have occurred at those points where the initial conditions were conducive to it. Most important, chaotic inflation was able to explain the large-scale structure of the universe.

The story does not end there, though. In 1986, Linde realized that chaotic inflation would be eternal—that is, some parts of space-time would always be inflating. That's because as the scalar field slowly rolls down a slope, in some regions the field gets a quantum-mechanical kick that pushes it back up the slope. In these regions, inflation takes off with greater ferocity, causing them to expand even faster than the surrounding regions. And within those patches of spacetime, the same thing can happen again, and again, and again. The theory—now called chaotic eternal inflation, or simply eternal inflation—predicts a mind-numbingly large and diverse universe. According to the theory, the process ended in our patch of spacetime and gave rise to our universe, the observable part of which is about 93 billion light-years across. Other parts of spacetime might still be inflating. The universe, then, is a fractal patchwork of regions, each with different properties. When and how inflation ends in a particular patch has to do with the slope the scalar field rolls down, and this also dictates the particular region's properties, such as the value for the energy density of its vacuum, or the exact nature and types of particles in it. Our universe is a part of this patchwork of regions of spacetime, or multiverse.

One of the first confirmations of this theory could come from the Planck satellite. Linde's version of inflation requires a relatively high energy density for the scalar field. Such a field would have generated gravitational waves strong enough to be detected by Planck. If the probe does see signs of these waves in the CMB, it would be a huge boost for eternal inflation and its prediction of a multiverse.

Not that Linde needs convincing. By the mid-1980s, he had realized that his theory more than hinted at a multiverse and also provided the necessary grounding for the controversial idea of the anthropic principle, which says that our universe has the properties it does because we are here to say so—since if it were any different it would possibly be hostile to life, and we wouldn't be around to ask questions. To avoid the thorny question of why the universe is just right for life, the anthropic principle works best when the properties of the universe are an accident of nature. This is where eternal inflation helps. If, as the theory suggests, there are many universes, each with its own randomly determined properties, then there is a small probability for a life-producing universe like ours to emerge. Alan Guth once called inflation "the ultimate free lunch," given that all matter and energy emerges from the initial false vacuum. Linde extended this metaphor, calling the inflationary universe "the only lunch at which all possible dishes are available."

At the time, physicists weren't very keen on the anthropic principle. Linde suspects that Western society, with its monotheistic underpinnings, has a visceral dislike for the idea. "The standard monotheistic tradition tells you that there is one God, one universe, and one set of laws. 'And [God says] if you do not obey me, if you do not believe in me, then I'll punish you and all your relatives,'" said Linde, as we sat in the living room of his Stanford home not long ago. A Meade telescope stood on a tripod in a corner of the room and a poster titled "The Origin and the Fate of the Universe" hung on the wall. The anthropic principle, however, got the cold shoulder even in the godless Soviet Union. In 1990, Linde moved to the United States, hoping that freethinking Americans would embrace the notion. But they, too, were "absolutely, adamantly against it."

Linde was ahead of his time, not just in his embrace of the anthropic principle but also in his anticipation of another major development: the "landscape" of string theory (see chapter 6). In a paper presented at a 1982 Nuffield Workshop in Cambridge, U.K., he commented on the ongoing efforts to deal with the extra dimensions of string theory. Physicists were looking at ways in which six of the ten dimensions of spacetime could be curled up, giving rise to the four-dimensional spacetime familiar to us. Linde suggested that if the extra dimensions could be scrunched up, then there "will be infinitely many mini-universes with [four dimensions] in which intelligent life can exist."

String theorists were keen on discovering a unique way of curling up the extra dimensions that would lead them to a theory that described *our* universe—ours and ours alone. But all indications were that there were myriad ways of scrunching spacetime, and all these "compactifications" seemed equally likely. By 1986, Linde was arguing that this was string theory's strength, not its weakness: "[The] enormously large number of possible types of compactification which exist [in string theory] should be considered not as a difficulty but as a virtue . . . since it increases the probability of the existence of mini-universes in which life of our type may appear."

The resistance to the anthropic principle began to weaken the following year, when Steven Weinberg published a landmark paper that provided yet another rationale for it. Weinberg had started looking for a natural explanation for the energy of the vacuum of space. At the time, most cosmologists thought that the likely value for vacuum energy was zero. The theories of particle physics, however, were showing that, far from being zero, the vacuum energy had to be much, much larger. To reconcile this mammoth gulf, Weinberg asked an anthropocentric question: What would be the value of vacuum energy that would be consistent with the evolution of a universe in which intelligent beings would arise to measure it? The answer was surprising. Weinberg showed that the energy density of the vacuum of space would have to be just a few times greater than the matter density of the universe. If it were much different, then stars and galaxies would not have formed and (therefore) life would not have arisen to inquire about the universe. Weinberg argued that if his calculations were correct, astronomers would soon be able to measure the vacuum energy density. And in 1998, they did. They called it dark energy, and its density was incredibly close to Weinberg's prediction (which had been refined subsequent to the 1987 paper). In essence, what the anthropic principle is saying is that the value of dark energy—or the cosmological constant, as it is also called (see chapter 4)—is best explained as an accident of the initial conditions of the universe. At least for now, it seems impossible to explain it from first principles. But bowing to the anthropic principle comes with a big caveat. It "makes sense if, and only if, you have this variety of universes," Weinberg remarked during my visit with him in Austin.

The idea of the multiverse has gained so much currency—especially as a way of explaining why some parameters of our universe appear to be fine-tuned for life—that even passionate proponents of a single universe are coming around. One of them is Brian Greene, a mathematical physicist at Columbia University. Greene's 1999 bestseller, *The Elegant Universe,* was an eloquent argument in favor of string theory and its promise to explain our universe from first principles. "A few years ago, I was a very strong believer in a single universe, whose laws and properties would be uniquely determined from some deep understanding of the laws of physics," he said when I reached him at his Columbia office. "In recent years, I have personally undergone a sort of transformation, where I am very warm to this possibility of there being many universes and the explanation is simply that we are in the one where we can survive."

Both string theory and eternal inflation are making a case for a multiverse—but until recently the two theories did not sit well together. The reason has to do with string theory's central challenge: how to turn ten dimensions of spacetime into four. The standard way of compactifying these extra dimensions is by using mathematical forms called Calabi-Yau manifolds, in which six of the ten dimensions are curled up tightly enough to be unnoticeable in our universe. Each one of these compactifications gives rise to a different vacuum of spacetime. In 2000, Joseph Polchinski and Raphael Bousso showed that certain properties of these manifolds—known as fluxes—could also vary, giving rise to even larger numbers of spacetimes. But there was one serious problem: None of these spacetimes were stable. The extra dimensions either curled up or blew up beyond acceptable limits—and one of the reasons they did so was because of the introduction of the fields needed for inflation. That problem was solved in 2003, when Shamit Kachru, Renata Kallosh, Linde, and Sandip Trivedi figured out a way to make the extra dimensions stable. The KKLT mechanism, as it is now called, was a landmark achievement in string theory. The quartet joined forces with other physicists to show how inflation could occur within the context of string theory—and it basically has to do with the mathematically nontrivial task of finding the right kind of scalar fields that could give rise to inflation. Michael Douglas, then at Rutgers University, and Leonard Susskind of Stanford put all the pieces together, and Susskind argued for the landscape of string theory, which predicts a staggering 10^{500} or more different universes.

The inflationary era in the context of the string-theory landscape is tumultuous and frantic. Imagine a patch of spacetime that has randomly determined properties. If the initial conditions are good for inflation, the process begins. The scalar field responsible for inflation slowly starts rolling down a slope. But then something incomprehensible happens. An inflating universe tunnels into another part of the landscape. It's as if one universe had morphed into an entirely different one.

To appreciate this bizarre process, it's useful to imagine the landscape as a terrain of hills and valleys. Each valley represents some field or parameter that is falling to a point of local stability. Because string theory has hundreds of such fields and parameters, the terrain is actually hyperdimensional, with valleys representing the most energetically stable values for each parameter. As a universe is inflating and its scalar field is rolling down to the base of some valley, it suddenly tunnels through adjacent hills and ends up on another slope elsewhere in the landscape. This is not as outrageous as it sounds. In quantum mechanics, a system will go from a higher energy state to a lower one—as long as it's not forbidden by some conservation law—even if there are barriers in its path. It does this via a phenomenon called quantum tunneling. Electrons, for instance, will do this routinely, in laboratory conditions. "That's the analog of what's happening here, except it's not a particle that's tunneling from one state to another, it's an entire universe that's tunneling from one state to another," said Greene.

In the case of string theory, it turns out that the hyperdimensional complexity of the landscape makes tunneling extremely likely and in timescales shorter than the time it takes for inflation to complete. So, essentially, an infant universe ends up exploring the landscape, hopping from place to place, until finally inflation ends and space-time—at least for the moment—finds itself in a stable place. And since inflation itself is eternal, patches of this spacetime would still be inflating, each its own universe. "There are all sorts of universes undergoing various trajectories, and we are simply one—our trajectory has landed us here, in some stable or pseudostable [state]," Greene said. "Who knows what's going to happen tomorrow, or a hundred billion years from now? Conceivably, we are in some pseudostable configuration that is momentarily hospitable to life. That moment might be very long, so I don't want to alarm anybody."

Earlier in his career, when string theory's promise of a unique solution seemed likely to be fulfilled, Greene spent a lot of time and effort trying to figure out how experiments could make contact with theory. He was specifically looking for observable phenomena that would fall out of the various ways of compacting extra dimensions. The hope was that when string theory did pick a particular compactification that gave us our universe, he would be ready with some physics. "That's what I truly believed, and what many others believed," he said.

But the discovery of dark energy and the emergence of the string-theory landscape has changed everything. "So what do you do?" said Greene. "Either you say that something is fundamentally wrong, or you say, 'What was really wrong was my prejudice that there should be one [universe] emerging from the theory.'" For Greene, it was time to question his prejudice. "You smack yourself in the head and say, 'Ah, maybe the universe is trying to tell me something.' The math is showing me many universes, the observations are most naturally explained with many universes, let me change my prejudice."

•   •   •

Quantum tunneling through the landscape has experimental consequences, and Planck is our best hope for testing them. If our universe emerged after a series of tunneling events, then some models of the universe constructed using string theory predict that space-time will have negative curvature, making our universe an open universe (whereas positive curvature leads to a closed universe). These models assume that just enough inflation occurred to explain certain properties of our universe (such as its homogeneity) but stopped short of making it absolutely flat. An open universe—that is, one with negative curvature—is one in which the cosmological parameter Omega, the ratio of the density of the universe to the critical density needed to make it flat, is less than 1. Current measurements are showing that Omega is equal to 1, meaning the universe is flat (see chapter 4). "But the uncertainties are such that it could also be either a slightly closed universe or a slightly open universe," said Linde. "It is difficult to obtain a closed universe from string theory—difficult but not completely impossible. So this is one of the things that some people are thinking about right now."

Planck and continuing observations by WMAP will help lessen the uncertainties in the measurements of Omega. As noted, another key measurement will be the detection of the signature of inflationary gravitational waves in the CMB. If Planck—or, indeed, any of the ground-based radio telescopes in Antarctica or Chile—see the B-mode polarization, it will suggest that gravitational waves were powerful enough during inflation to have polarized the CMB photons, in turn fixing the energy scale of inflation. The sensitivity of today's instruments is such that if they were to see this effect, it would mean that inflation necessarily occurred at high energy scales. Until recently, most string-theory models predicted that inflation did *not* occur at high energy scales, so they would not have produced gravitational waves that could be detected with Planck or WMAP. Then, in 2008, Eva Silverstein of Stanford and her colleagues identified at least two models of eternal inflation using string theory in which the energy scales are high enough to have observable consequences. No wonder, then, that cosmologists are eagerly awaiting the detection or nondetection of primordial gravitational waves in the next few years.

Oddly enough, if the CMB experiments find evidence of gravitational waves, it will have implications for the search for supersymmetry at the Large Hadron Collider. A high-energy scale for inflation means that the mass of the gravitino—the partner particle of the graviton in supersymmetric theories, the carrier of the force of gravity—will be more than a thousand times the mass of the proton. While the LHC is not looking for gravitons, it is looking for other supersymmetric particles, and a very heavy graviton would put those particles out of reach. The energy scale at which supersymmetry appears would be well beyond the range of the LHC.

On the other hand, if the LHC does see signs of supersymmetry at the TeV scale, then Planck probably won't see gravitational waves—the mass of gravitons would likely be too small to have generated waves detectable by current CMB experiments. So cosmologists are already proposing another experiment called CMBPol (for "Polarization")—a more specialized and sensitive version of Planck. But if the LHC detects signs of supersymmetry *and* Planck finds gravitational waves, the theorists will have to scramble to explain it. Nothing helps focus the mind better than conflicting experimental results.

There is yet another way in which Planck could help constrain models of inflation. It has to do with something odd in the WMAP data. The simplest models of inflation predict that the amplitudes of the temperature anisotropies in the CMB should be spread across the sky in what's called a Gaussian distribution—a bell-shaped curve. But five years' worth of WMAP data hint that the distribution might be somewhat non-Gaussian. It's a controversial claim, because this is an extremely difficult measurement, equivalent to looking for variations within the variations of temperature. For theoreticians, a non-Gaussian distribution is difficult to generate and requires complex models of inflation, with multiple interacting scalar fields (a fact that does not faze some string theorists, for their models are flush with multiple fields). "This non-Gaussianity is much larger than what is predicted by the simplest versions of inflationary models," said Linde. "But the question is whether they have *really* discovered it or not. Right now, there is a general state of confusion, and we need better measurements. Planck in particular will do that. That is one of its goals and one of our dreams."

Planck, along with the LHC, the dark-matter experiments, neutrino telescopes, and the dark-energy experiments, all promise a great deal over the next decade. The puzzle of dark matter will most likely be solved in the coming years via a combination of three techniques: First, the indirect detection of particles such as electrons, positrons, protons, antiprotons, and neutrinos that should be produced when large accumulations of dark-matter particles annihilate. Already, a balloon-based experiment called ATIC (for Advanced Thin Ionization Calorimeter) and a space probe named PAMELA (Payload for Antimatter Matter Exploration and Light-nuclei Astrophysics) have provided tantalizing evidence of an excess of electrons and positrons in space that could be due to the annihilation of dark matter. Second, direct detection of dark matter using experiments like CDMS II inside the Soudan Mine. Third, the production of dark-matter particles at the LHC. When all three techniques converge and start pointing toward the same physics, we will finally know the nature of dark matter.

Dark energy will prove a much tougher problem to solve. In the years to come, thanks to telescopes like Planck and the Square Kilometre Array, we might know if the density of dark energy has varied over time. Is dark energy Einstein's cosmological constant, or is it something more exotic? Physicists will still be left with the question of why the dark-energy density is what it is. Which will bring us back to the string-theory landscape, the multiverse, and the anthropic principle. Physicists will have to step gingerly through this theoretical minefield, treading only on those parts that experimentalists say are safe. The terrain will be illuminated by any evidence of quantum gravity from IceCube, or hints of supersymmetry from the LHC, or signs of gravitational waves in the CMB.

It's both humbling and heady that theoretical concepts as grand as the landscape and the multiverse are within reach of experiments—even if barely. Heady because no one, even a decade ago, would have imagined that such tests were possible. Humbling because no matter how immense the theoretical edifice, it must ultimately bow to experimental verification. And experiments will chip away at any edifice until the underlying form of the right theory begins to emerge. We may never be able to detect the existence of other universes, but that does not deter experimentalists. As Weinberg said, "The important thing is not whether you can observe every ingredient in a theory, the important thing is whether you can observe enough of the consequences of the theory to test it and confirm that the theory is right."

For now, we can only watch and wait, while physicists listen for the whispers from other universes.

EPILOGUE

THE HIMALAYAN VALLEY was eerily quiet. There was no sound of industry—no factory, no generators, no machines. Not even the *putt-putt* of the ubiquitous Indian scooters. A few wild horses grazed on the meager grass—more brown than green—and a handful of men and women, their backs bent, tilled the land for their summer crop of peas and barley. Small clusters of mud-brick houses with lime-washed walls dotted the valley. The only trees were bare, scrawny willows that had yet to shake off the effects of a harsh winter. The narrow Hanle River, fed by Himalayan snowmelt, flowed near the base of the mountains to the south. The river must once have raged with ravenous energy to carve out this valley, entering from the southwest and exiting through a narrow opening in the east. Thousands of years ago, when the opening was accidentally dammed by falling rock, it turned the valley into a freshwater lake. Eventually the source dried up and the lake turned brackish. Today, all that's left is a flat, spongy, marshlike lakebed about 3 miles across and a handful of hills that the river could not erode.

I was standing on one of those hills, named Mount Saraswati, after the Hindu goddess of learning. Only here can a peak 4,517 meters high be called a hill; the valley floor is itself at an altitude of 4,240 meters. I was deep inside the mountainous region of Ladakh, in northern India, in a place called the Hanle Valley, a small corner of the Tibetan Plateau. Mount Saraswati is near the center of the valley and is surrounded by mountains up to 6,000 meters high. To the east are peaks that look like something a child might draw: They rise in waves from the lakebed, with sensuous flanks and ridges, each peak a different shade of brown, some tinged with green and red. The highest mountains have spiky, snow-dusted tops, in sharp contrast to the bare, rounded ones beneath. Behind them flows the Indus River—part of the military line of control between India and China.

Perhaps surprisingly, Hanle is poised to become one of cosmology's most sought-after sites. Indian astronomers have just begun studying the quality of the seeing from atop the hills in this region. Ladakh itself opened up to the world only a few decades ago; until recently, the only way in was over treacherous mountain roads that crossed the Greater Himalayas. One route goes over some of the highest motorable passes in the world, two of which are above 5,000 meters; the route is open for less than four months each year. I cheated, flying over the Himalayas into Leh, Ladakh's largest town. But going directly from sea level to an altitude of 3,500 meters has its perils. At the guesthouse run by the Indian Astronomical Observatory, I was ordered to rest. The best way to acclimatize to the oxygen-poor atmosphere, they said, was to drink liters of water and sleep for a day and a half. Any serious exertion on the first day was forbidden. Tourists usually ignore this dictate, only to find themselves in the local hospital. I was asked not to take walks or even a bath. I slept fitfully that night. The air was extremely dry, parching my nose, mouth, and throat. An altitude-induced headache continued through the night, settling like a clamp around my forehead. I felt unreasonably tired. Even pulling the heavy woolen blanket over me to keep out the biting cold left me gasping.

Three days in Leh and I was ready to head to higher altitudes. Tushar Prabhu, an astronomer from the Indian Institute of Astrophysics in Bangalore and the project manager for the Hanle observatory, was my host. It was a seven-hour drive to Hanle, on a road that followed the Indus River much of the way. In Ladakh, nature revels in naked glory. Tectonic forces over the past 50 million years have shaped the great mountain ranges of this region. The Himalayas formed when the Indian Plate, drifting north across the ancient

Tethys Ocean, slammed into the Eurasian Plate and started slipping beneath it. Evidence of that violence was all around us. The road sometimes ran along gigantic granite cliffs of the Ladakh Range that had been forged in the heat of the collision. Far beneath them, the Indus was an aquamarine stream winding its way through a gorge. Across the gorge were the equally massive cliffs of the Zanskar Range: layers and layers of seafloor sediment that had been uplifted in the smashup and now formed immense slabs of sandstone stacked at impossibly steep angles.

Along the road were the occasional (and extremely optimistic) crash barriers: piles of stones about two feet high, a foot wide, and three feet long. Stacking stones seems to be an art form in Ladakh. Everything from fences to the walls of houses, especially in villages, is made of stones arranged with such skill that they stand solid without mortar to bind them. *Mani* walls are everywhere. These are like the stone crash barriers but bigger, and the flat stones on top are exquisitely engraved with prayers. The wind blowing over the stones is said to carry the prayers to the gods. This is an intensely Buddhist land. Ubiquitous prayer wheels are spun to send invocations into the air. Everywhere you look, there flutter multicolored flags printed with the sacred words *Om Mani Padme Hum* (a phrase said to contain the "essential meaning of all eighty-four thousand sections of the Buddha's teachings"). Even the observatory that I was traveling to is linked to Buddhist lore. The village of Hanle is dominated by a four-hundred-year-old monastery. When the astronomers were scouting the region, they found a hill that offered the best seeing within easy reach of the unpaved road that reached the village. They decided to flatten its top, but they came upon an old Buddhist structure near the summit. It was an ancient enclosure of slate stones about a foot high and two feet across that contained mud statues of gods. The astronomers approached the *Rinpoche*, the head of the Hanle monastery, for advice on what to do with it. He said they could dismantle the structure and build their telescopes provided they built a small stupa (a more permanent reliquary) elsewhere on the hill. "We discovered how open-minded and progressive he was," said Prabhu.

The Himalayas are the youngest and the highest of the great mountain ranges in the world. And the Hanle observatory, nestled in these mountains, is still in its infancy. Normally, the monsoon rains of the Indian subcontinent render much of the land unsuitable for serious astronomy for months at a time. But the Hanle Valley lies north of the Greater Himalayas and its wall of 8,000-meter peaks, which block the monsoons, leaving the Tibetan Plateau dry. So dry, in fact, that even the banks of the Indus were almost devoid of vegetation. The area is also relatively free of aerosols—fine particles of dust—that act as nuclei for water vapor to condense, producing fog and clouds.

Mount Saraswati hosts a 2-meter-class (80-inch) telescope reminiscent of the 100-inch at Mount Wilson. Today the small telescope at Hanle is part of an international collaboration called COSMOGRAIL (for COSmological MOnitoring of GRAvItational Lenses), which is using telescopes spread around the world to study gravitational lenses to accurately determine the value of the Hubble constant. Astronomers have continued to refine the value of this constant ever since Hubble measured it from Mount Wilson, for it remains a crucial ingredient in any studies of the expansion of the universe. Hanle is uniquely positioned to help out. There are no other observatories in the Eastern Hemisphere where the conditions match or exceed those of observatories in the Chilean Andes or on Mauna Kea. COSMOGRAIL relies on uninterrupted monitoring of gravitational lenses to pick up on minute temporal variations in the behavior of the lenses; this is what allows astronomers to hone the measurement of the Hubble constant. To help them do so, telescopes hand off the observation from one hemisphere to another as day follows night.

Continuous monitoring of transient sources such as supernovae and gamma-ray bursts are also becoming increasingly important. For example, dark-energy studies will rely on a detailed understanding of supernovae, which in turn will rely on knowing what's going on inside stars before they explode. One way to study the inside of stars is by looking at their surface oscillations. This emerging field of astro-

seismology requires close and constant observation. Take our closest star, the sun. Helioseismology has allowed scientists to study the
sun's oscillations and get a firm grip on the nuclear reactions that occur deep within it. This understanding played a crucial role in solving
the mystery of the missing solar neutrinos. When neutrino detectors
were monitoring the sun, they kept finding a deficit of neutrinos. The
problem was so vexing that some physicists thought they might have
to revise their models of nuclear reactions inside the sun. However,
our knowledge of the sun was so good that the problem undeniably
pointed toward a lack of understanding of neutrinos. As we saw in
chapter 3, eventually theoreticians realized that neutrinos were oscillating, or changing form, on their way from the sun to Earth, and
this led to one of the first serious experimental challenges to the
standard model of particle physics. Challenging the standard model
of cosmology will require an equally in-depth understanding of other
stars, astrophysical objects, and processes.

Despite its ideal location in the Eastern Hemisphere, Hanle is
hostile to astronomers. Temperatures can drop down to –22°F in the
winter, the season when the night sky is especially clear, easily bet-
tering the view from Mauna Kea. So astronomers have focused on
developing a robotic telescope that can be controlled from the moderate climes of Bangalore, in southern India. As astronomers venture
higher and higher and into ever more remote regions to build observatories, these robotic telescopes will be the key. Another robotic
telescope sits in Dome C in East Antarctica, 10,607 feet above sea
level. The dome is higher on the Antarctic Plateau than the South
Pole, and temperatures can plummet to –112°F in midwinter, the
only time when it's dark enough to observe the skies. But it is an ex-
ceptional place for astronomy. The dry and stable air above the Antarctic Plateau means that a telescope there can outperform mid-lat-
itude telescopes two or three times its size. Astronomers have built
a small robotic telescope at Dome C as a precursor to a bigger one;
in July 2007 it studied two stars, a binary and a variable, for 243
hours straight, the longest ever continuous observation of particular
stars.

Soon, newer observatories like Hanle will likely host 10-meter-class telescopes and become part of a significant shift. Such telescopes will replace today's 4-meter-class telescopes as the workhorses of the future, leaving the deeper and finer explorations of the cosmos to the next generation of even bigger instruments: the 24.5-meter Giant Magellan Telescope, the Thirty Meter Telescope (being designed by the same group that built the Keck), and the 42-meter European Extremely Large Telescope, all of which are in various stages of development. These giants will take on the task of studying the formation of the very first stars and galaxies and the evolution of large-scale structure from its earliest moments to the present time. They will help determine whether the so-called fundamental constants have changed with time, probe the structure of spacetime, and possibly lead us toward a theory of quantum gravity. Today's ground-based telescopes have worked well with the Hubble Space Telescope, and tomorrow's giants will have their own partner: NASA's 6.5-meter James Webb Space Telescope, due to be launched in 2014.

Optical telescopes are not the only instruments in the works. One of the most anticipated observatories is the Laser Interferometer Space Antenna (LISA), which will directly detect gravitational waves as they ripple through spacetime. Until now, the existence of gravitational waves has only been inferred, mainly from the study of binary pulsars. As noted in the preceding chapter, measurements of the polarization of the cosmic microwave background could also reveal the presence of gravitational waves, but only indirectly. To actually "sense" these waves, LISA will use three small spacecraft flying in formation at the vertices of an equilateral triangle. The spacecraft will be 5 million kilometers from each other and the formation will orbit the sun at a distance of 50 million kilometers from Earth. Precision optics and lasers will help the craft keep track of their relative positions to within tens of nanometers. If a gravitational wave were to pass through, it would squeeze and expand spacetime, changing the distance between the craft in ways that could be picked up by LISA. Detectors built on Earth, such as the Laser Interferometer Gravitational-wave Observatory (LIGO), are also searching for such waves, but they are not yet sensitive enough for the biggest prize: gravitational waves from an infant universe. Just as the microwave background permeates the cosmos, so gravitational waves generated just after the big bang are thought to suffuse spacetime. LISA aims to detect them.

Besides LISA, a slew of space-based instruments are being planned to study dark energy, as part of NASA's Joint Dark Energy Mission (JDEM). But space is an expensive place to do science. Cosmology's interests are well served by identifying and maintaining pristine sites on Earth that can complement space. Dome C and Hanle represent the next generation of such sites.

Two days after we reached the Hanle observatory's base camp, I decided to climb to the summit of Mount Saraswati. Sonam, a Ladhaki engineer, suggested I proceed slowly in case I hadn't yet adjusted to the altitude. I smirked inwardly — the summit did not look particularly far. Within minutes, though, I knew he was right. I was drawing deep breaths but my lungs refused to respond. Tightness settled around my chest. I could barely take two steps before having to stop. I literally inched my way to the top. At the summit, prayer flags flapped in the steady wind. The 2-meter telescope shares the summit with three stupas, one built by the astronomers and two more built by the locals for the Dalai Lama, who has twice visited the observatory.

From the top, I could see right across the valley. All around were imposing snowcapped mountains. To the east, in front of mountains that looked like a child's drawing, was the many-storeyed Hanle monastery. Its whitewashed walls rose off the steep slopes of a greenish, craggy hill, making its perch look precarious, but it had lasted four centuries and would surely last many more. I was struck by the juxtaposition: Two outposts, one ancient and one modern, stood watch over the valley. Both hope to benefit humanity, one by focusing energy on the human mind, the other by gazing toward the skies. Both, in a way, are trying to illuminate our future by studying our past.

The timeless culture of Ladakh is changing, under assault from tourists. Hanle, however, is deep within Ladakh and relatively safe from polluting influences. But the same cannot be said for some of the other sites I visited. The Russians are building an oil pipeline not far from Lake Baikal's northern shores. The government has promised to keep the pipeline out of the lake's drainage areas, but environmentalists remain wary. In the early 1960s, the government ignored their pleas and went ahead and built that paper and pulp mill in Baikalsk. The mill dumped chemicals into the lake for years before shutting down in late 2008, a victim of the global recession.

These local threats are dwarfed by something that does not respect geographical boundaries: climate change. Over the past fifteen years, the frequency of so-called bad winds has increased over Cerro Paranal in Chile, leading to turbulence near the summit, which degrades the quality of seeing. These winds have been attributed to global warming. If our planet's temperature continues to rise as feared, eventually the ice sheets and ice shelves of Antarctica will feel the heat. If the ice were to melt, we would lose our best natural asset for detecting neutrinos, and with it our ability to probe the hidden recesses of our universe.

On Mount Saraswati, I became aware of the deep silence enveloping me. The monks had built their monastery here for a reason. Now, the astronomers were taking advantage of the intense quiet to find answers to profound questions. It was abundantly clear, standing in Hanle, as it had been in places like the South Pole, Lake Baikal, Paranal, and the Karoo, that the natural calm of these places is what makes them ideal for cosmology. We need to protect them—just as we protect our inner peace as individuals. If we pollute them, we will destroy our best chance of deciphering our own beginnings, of understanding ourselves.

APPENDIX I

The Standard Model of Particle Physics

There are four known fundamental forces of nature: electromagnetism, the weak nuclear force, the strong nuclear force, and gravity. The standard model of particle physics is the theory of matter and all the fundamental forces except gravity.

The model slices and dices its set of fundamental particles in a number of ways. The broadest categories are bosons (the carriers of forces) and fermions (which make up matter):

- Bosons include the photon (electromagnetic force), the W and the Z particles (weak nuclear force), and the eight types of gluons (strong nuclear force). The photon and gluons are massless, but the W (at 80 GeV, about eighty times the mass of a proton) and the Z (91 GeV) are some of the most massive particles ever discovered.

- Fermions are subdivided into leptons and quarks:

  - Leptons come in three groups: the electron and the electron neutrino; the muon and the muon neutrino; and the tau and the tau neutrino.

  - Quarks also come in three groups (which are aligned with the three groups of leptons): Up and Down; Charm and Strange; and Top and Bottom. The standard model shows how various combinations of three of these six quarks gives rise to particles called baryons, as well as other particles composed of different combinations of quarks. Protons (Up, Up, Down) and neutrons (Down, Down, Up) are the most familiar baryons. There are many others, but they are unstable and decay instantly. So an atom of normal matter is made of leptons (electrons) and baryons (protons and neutrons).

The standard model, while extremely well tested, is not perfect. For one, it doesn't include gravity. Also, the three fundamental forces described by the model do not unify at higher energies, and physicists have strong reason to think they should. An idea called supersymmetry (SUSY) extends the standard model by positing a bosonic partner for every fermion and a fermionic partner for every boson. In some theories, such as string theory, which incorporate supersymmetry, gravity gets its own boson: the graviton. Also, the forces unify accurately at higher energies. The standard model has been proved experimentally, but there is no evidence as yet that the universe is supersymmetric.

APPENDIX II

From the Big Bang to Now:
The Standard Model of Cosmology

At the big bang, the universe is an intensely hot and dense speck of spacetime. It begins expanding and cooling down. From the birth of the universe to 10^{-43} seconds later—a period called the Planck epoch—the four fundamental forces of nature are unified as one. At the end of this epoch, the symmetry between gravity and the other forces is broken. This begins the next phase, the grand unification epoch, which lasts until 10^{-36} seconds after the big bang, at the end of which the strong nuclear force separates from the electroweak force. This transition triggers inflation, a brief and violent episode that lasts just fractions of a second but during which the universe expands exponentially, becoming orders of magnitude bigger. When inflation ends, the universe reheats, giving rise to radiation and elementary particles such as quarks and electrons, and also particles of dark matter. By then, inflation has stretched the quantum fluctuations of the infant universe and turned them into small variations in the density of matter and energy, which eventually become the seeds for the large-scale structure of the universe. Homogeneous apart from these variations, the universe returns to the more sedate rate of expansion that existed prior to inflation.

At the end of the electroweak epoch, about 10^{-12} seconds after the big bang, even electromagnetism and the weak nuclear force part ways. The universe now has four distinct fundamental forces (see Appendix I). Also, a not-so-well-understood process known as baryogenesis has given rise to more matter than antimatter, leading to protons and neutrons. About three minutes after the big bang, the universe cools enough to form the nuclei of hydrogen and helium and a few other light elements. Still, the universe is hot enough that it continues to be a plasma of radiation, electrons, and light nuclei. Then, about 370,000 years after the big bang, when temperatures have dropped substantially, the electrons combine with the nuclei to form atoms of neutral hydrogen and helium, among others. This is the epoch of recombination. The photons of the cosmic microwave background are released. A few hundred million years later, the first stars and galaxies start to form. The universe keeps expanding, but its expansion rate is now being slowed due to the pull of gravity. Then, a few billion years after the big bang, the cumulative energy of the vacuum of the expanding spacetime—now known as dark energy—becomes strong enough to start repulsing gravity. The expansion of the universe, instead of slowing down, starts to accelerate.

This is the standard model of cosmology, parts of which remain speculative and untestable, given the current state of both theory and technology (particularly the Planck and grand unified epochs). This model is known as Λ-CDM, where Λ (lambda) represents dark energy and CDM stands for cold dark matter, which is dark matter that is composed of massive, slow-moving particles. Cosmologists do not know the exact nature of either dark energy or dark matter, but they have established some parameters of the model by measuring the cosmic microwave background and analyzing distant and nearby supernovae as well as the large-scale structure of galaxies. These experiments tell us that the universe is composed of dark energy (~73 percent), dark matter (~23 percent), and normal matter (~4 percent). The parameter Ω (Omega), which is the ratio of the density of the universe to the critical density needed to make it flat, is about 1. All evidence so far suggests that we are living in a flat universe.

NOTES

PAGE   *Epigraph*

xiii   A. L. Basham, *The Wonder That Was India* (New York: Grove Press, 1954), 248.

Prologue

3   *"The great mystery"*: http://www.newscientist.com/blog/space/2007/02/physicists-debate-nature-of-space-time.html.

4   *"have enough time"*: Ibid.
    "I think you can": Ibid.
    "All I can say": Ibid.

1. Monks and Astronomers

9   *Five hundred schoolgirls*: Isaacson, *Einstein: His Life and Universe*, 372.
    "He is not": *New York Times*, Jan. 1, 1931.

10   *To gather the wood*: Sandage, *The Mount Wilson Observatory*, 14.
    Benjamin Wilson: Ibid., 15.

11   *as he lay down*: Starr, *The Dream Endures*, 73.
    "he who would launch": Sandage, 170.
    He had been inspired: Starr, 77.

12   *"elite fellowship"*: Ibid., 78.
    "the specifics of optics": Ibid.
    "the profound question": Ibid.

13   *"It is certainly a wonderful"*: Astronomical Society of the Pacific, *The Adolfo Stahl Lectures in Astronomy* (San Francisco: Stanford Univ. Press, 1919).
    But first, the observatory: Schindler, "An Amazing Tale."
    "faint circular mark": Ibid.

14   *"Uncle John"*: Aitken, "John Alfred Brashear."
    "We have loved the stars": Ibid.

15   *When Slipher announced*: Bertotti et al., eds., *Modern Cosmology in Retrospect*, 262.

19   *"This was the astronomical"*: Sandage, 178.

"His face grotesquely": Christianson, "Edwin Hubble: Reluctant Cosmologist."

"stretch[ing] out into nothing": Ibid.

"The 'dark-Moon types'": Sandage, 162.

20   *"Astronomy is something"*: Singh, *Big Bang*, 218.

"His tall vigorous figure": Osterbrock et al., "Self-Made Cosmologist: The Education of Edwin Hubble."

Night assistants, who really run: Sandage, 188.

21   *"making him history's"*: Ibid., 192.

"lack of concentration": Singh, 204.

Convinced that women: Ibid.

22   *She was studying:* Johnson, *Miss Leavitt's Stars*, 37.

"I feel seriously inclined": Ibid., 118.

the letter never reached: Ibid.

23   *"faint praise"*: Ibid., 119.

24   *"sleek Pierce-Arrow"*: Isaacson, 354.

"to everyone's alarm": Christianson, *Edwin Hubble: Mariner of the Nebulae*, 206.

"Well, my husband": Isaacson, 354.

"European Hubble": Egret and Heck, eds., *Harmonizing Cosmic Distance Scales in a Post-HIPPARCOS Era*.

25   *"midnight lunch shack"*: Sandage, 191.

"As late as 1955": Ibid.

26   *"two pieces of bread"*: Ibid.

"Both the night": Ibid., 181.

27   *"high table at Oxford"*: Ibid., 525.

"almost pathologically shy": Ibid.

"mountains which seem": Henson and Usner, *The Natural History of Big Sur*, 272.

28   *"To know that what"*: Einstein, "The Merging of Spirit and Science," quoted in Hale, *Christ and the Universe*, v.

2. The Experiment That Detects Nothing

36   *"When I started to create"*: "Deconstruction: Soudan Mural," *Symmetry Magazine* 2:4, May 2005.

37   *"6-foot-3-inch, 30-year-old"*: Harper, "Getting a Bang Out of Gamow."

38   *"The first day"*: Gamow, quoted in Harper.

40   *"There are reasons"*: Ostriker et al., "The Size and Mass of Galaxies, and the Mass of the Universe."

"By 1982, after a decade": Rubin, "Galaxy Dynamics and the Mass Density of the Universe."

42   *"Dear Hans"*: Harper et al., eds., "Reminiscences of George Gamow."

3. Little Neutral Ones

62   *"Dear radioactive"*: Sime, *Lise Meitner: A Life in Physics*, 107.

"desperate remedy": Ibid.

"I admit that": Ibid.

the two scientists announced: Cowan et al., "Detection of the Free Neutrino: A Confirmation."

63 *entire case of champagne:* "The Reines-Cowan Experiments: Detecting the Poltergeist," *Los Alamos Science* 25 (1997).

64 *"It was a moving"*: Murayama, "The Origin of Neutrino Mass."

4. The Paranal Light Quartet

83 *Water is an extremely:* http://www.oas.org/dsd/publications/Unit/oea 59e/ch12.htm.

the same trick: Yoon, "Clues to Redwoods' Mighty Growth Emerge in Fog."

86 *"It is not surprising"*: Singh, *Big Bang*, 349.

88 *"Boys, we have been scooped"*: Ibid., 432.

93 *Each mirror, made:* Dierickx, "The VLT Primary Mirrors: Mirror Production and Measured Performance."

The first of the four: "First VLT Mirror Cell and 8.2-m Dummy Mirror Arrive at Paranal," ESO Press Release, Nov. 17, 1997. http://www.eso .org/public/outreach/press-rel/pr-1997/phot-30-97.html.

100 *"joyous news"*: Coles, "Einstein, Eddington and the 1919 Eclipse."

"This is the most": Ibid.

"there is no great chance": Schneider and Falco, *Gravitational Lenses*, 4.

"much better chance than stars": Ibid., 5.

It, too, was discovered: John Huchra, e-mail correspondence with the author.

5. Fire, Rock, and Ice

109 *"distracting beauty"*: Bird, *The Hawaiian Archipelago*, Letter viii, 112.

"The cascades are most": Ibid.

111 *In the early 1960s:* "In Memoriam: Mitsuo Akiyama, 1920–2004," Institute for Astronomy, University of Hawaii, Newsletter, Summer 2004.

116 *Kain called Joe Calmes:* Sinsheimer, *The Strands of a Life*, 261.

124 *"the air was surrounded"*: National Park Service, U.S. Department of the Interior, "Wahi Kapu o Pele." http://www.nps.gov/havo/forteachers/ upload/havo_edprog_2007_4thgrade.pdf.

125 *"A white dog"*: U.S. Department of Commerce, Mauna Loa Observatory, Web Museum, "Mauna Loa's White Dog." http://www.mlo.noaa .gov/webmuseum/mlodog.html.

"the land trembles": "Wahi Kapu o Pele."

6. Three Thousand Eyes in the Karoo

136 *"primeval solitude and silence"*: Bryce, *Impressions of South Africa*, 55.

137 *The Boers settled:* "How the Boers Measured Land," *New York Times*, Jan. 18, 1880.

140   *"Enter Grote Reber"*: Reber, "A Play Entitled the Beginning of Radio Astronomy," 97.
      "Clearly, Henry Ford": Ibid., 99.
      "air navigation rules": Ibid., 103.
      "after school": Ibid.

144   *"I'm in love with this"*: Ananthaswamy, "Radio Telescope Offers Dishes to Savour."
      "They told us that": Ibid.

157   *Taylor and several colleagues measured:* "The Nobel Prize in Physics 1993." http://www.nobelprize.org/nobel_prizes/physics/laureates/1993/press.html.

7. Antimatter over Antarctica

168   *"Nowhere else on Earth"*: Spufford (ed.), *The Antarctic*, 4.
170   *"subzero stillness"*: Ibid., 84.
      "If we had been dressed": Cherry-Garrard, *The Worst Journey in the World,* vol I.
      "Beyond was the frozen": Ibid.
171   *"Now for the run home"*: Scott, *Scott's Last Expedition.*
172   *"We shall stick"*: Ibid.
      "Carrying [the rocks] along": Spufford, 8.
178   *"The tractor was gone"*: Belanger, *Deep Freeze,* 76.
181   *"If we accept the view"*: Dirac, "Theory of Electrons and Positrons."
185   *"perhaps somewhat selfishly"*: Glines, ed., *Polar Aviation,* 109.
      "It was not until": Ibid.
192   *"Are you all well?"* Cherry-Garrard, vol II.
      "The Polar Party died": Ibid.
      "Observation Hill was": Ibid.
      "I do not suppose": Ibid.

8. Einstein Meets Quantum Physics at the South Pole

195   *"It is a wonderful sight"*: Spufford, ed., *The Antarctic*, 59.
      "Scott for scientific method": Sir Peter Blake Trust, "Special Report One: The Antarctic Explorers." http://www.sirpeterblaketrust.org/blakexpeditions/special_report/3726.
196   *"People thought it a matter"*: Fridtjof Nansen, introduction to Amundsen, *The South Pole.*
198   *"No other moment"*: Amundsen, chap. xii.
199   *"Great God"*: Scott, *Scott's Last Expedition.*
200   *"Abandonment of the station"*: Unclassified memorandum, U.S. National Security Council, Mar. 9, 1996. http://www.fas.org/irp/offdocs/pdd26.htm.
214   *"Conrad (Gus) Shinn revved"*: Martin, *Hellbent for the Pole,* 126.
      "He called the stations": Ibid., 141.
220   *"And this was the thought"*: Edward Wilson, quoted in Turley, *The Voyages of Captain Scott,* 373.

9. The Heart of the Matter

223 *"It takes machines like"*: Cox, "A Journey to the Edge of Understanding."

225 *"Shiva rises from his rapture"*: Gannon, *Understanding Global Cultures*, 65.

226 *Civil engineers could not afford to wait*: The Challenges of LHC Civil Engineering, a film by M. Buhler-Broglin, CERN/AC, CERN ETT-EC-MM, 2003.

228 *It is the "binding energy"*: Frank Wilczek, "The Origin of Mass and the Feebleness of Gravity." http://mitworld.mit.edu/video/204/.

243 *"Throw deep"*: Kevles, *The Physicists*, xx.

246 *"boson named after me"*: Randerson, "Father of the God Particle."
     "The only reason": http://www.newscientist.com/blog/shortsharpscience/labels/quantum.html.
     An atheist: Ibid.
     "I shall open": Ibid.
     "I hope so": Ibid.

247 *"I have just enough"*: Weinberg, "Living in the Multiverse."

10. Whispers from Other Universes

248 *glittering stars of cosmology*: http://www.newscientist.com/blog/space/2007/02/precision-cosmology-with-planck.html.

249 *"What is your confidence"*: Ibid.
     "I'm surprised": Ibid.

252 *"It's a beauty"*: Ibid.

256 *"but COBE was bolted"*: Mather, "From the Big Bang to the Nobel Prize and Beyond."

257 *"There was just"*: Smoot and Davidson, *Wrinkles in Time*, 253.
     "If you're religious": Ibid., 289.

258 *"Matter tells spacetime"*: Wheeler and Ford, *Geons, Black Holes & Quantum Foam*, 235.

260 *"Listen Closely"*: Glanz, "Listen Closely: From Tiny Hum Came Big Bang."

262 *When it finally reached*: Linde, "The New Inflationary Universe Scenario."
     In 1983, he came: Linde, "Chaotic Inflation."

264 *"ultimate free lunch"*: Guth, quoted in Hawking, *A Brief History of Time*, 129.
     "the only lunch": Linde, "The New Inflationary Universe Scenario," reprinted in Gibbons et al., eds., *The Very Early Universe*, 245.

265 *"will be infinitely"*: Ibid.
     "[The] enormously large": Linde, "Eternally Existing Self-Reproducing Chaotic Inflationary Universe."
     a landmark paper: Weinberg, "Anthropic Bound on the Cosmological Constant."

267 *it's useful to imagine the landscape*: http://www.fqxi.org/community/articles/display/113.

270 *cosmologists are already*: Baumann et al., "CMBPol Mission Concept Study."

Epilogue

275   *"essential meaning of all"*: Khyentse, *The Heart Treasure of the Enlightened Ones*, 58.

280   *In the early 1960s:* Begley, "The Recession's Green Lining."
      Over the past fifteen years: Sarazin et al., "Seeing Is Believing: New Facts About the Evolution of Seeing on Paranal."

BIBLIOGRAPHY

Aitken, R. G. "John Alfred Brashear," *Pub. Astron. Soc. Pac.* 32:187, 175 (1920).

Alpher, R. A., H. Bethe, and G. Gamow. "The Origin of Chemical Elements," *Phys. Rev.* 73:7, 803–4 (1948).

Amundsen, Roald. *The South Pole: An Account of the Norwegian Antarctic Expedition in the "Fram," 1910–1912.* http://www.gutenberg.org/dirs/etext03/tsp1210h.htm.

Ananthaswamy, Anil. "Radio Telescope Offers Dishes to Savour," *New Scientist,* Feb. 19, 2005.

Baumann, Daniel, et al. "CMBPol Mission Concept Study: Probing Inflation with CMB Polarization." arXiv:astro-ph/0811.3919v2.

Begley, Sharon. "The Recession's Green Lining," *Newsweek,* Mar. 16, 2009.

Belanger, Dian Olson. *Deep Freeze: The United States, the International Geophysical Year, and the Origins of Antarctica's Age of Science* (Boulder, CO: Univ. of Colorado Press, 2006).

Bertotti, B., et al., eds. *Modern Cosmology in Retrospect* (Cambridge, U.K.: Cambridge Univ. Press, 1990).

Bird, Isabella L. *The Hawaiian Archipelago* (London: John Murray, 1875). http://ebooks.adelaide.edu.au/b/bird/isabella/hawaii.

Blake, C. A., et al. "Cosmology with the SKA," *New Astron. Rev.* 48:11–12, 1063–77 (2004).

Blegen, Theodore C., and Russell W. Fridley. *Minnesota: A History of the State,* 2d ed. (Minneapolis: Univ. of Minnesota Press, 1975).

Blumenthal, George R., et al. "Formation of Galaxies and Large-Scale Structure with Cold Dark Matter," *Nature* 311:517–25 (1984).

Bryce, James, 1st Viscount Bryce. *Impressions of South Africa* (London: Macmillan, 1899). gutenberg.org/etext/22323.

Casadei, Diego. "Searches for Cosmic Antimatter." arXiv:astro-ph/0405417v3.

Cherry-Garrard, Apsley. *The Worst Journey in the World,* vols. I & II. gutenberg.org/etext/14363.

Chongchitnan, Sirichai, and George Efstathiou. "Prospects for Direct Detection of Primordial Gravitational Waves." arXiv:astro-ph/0602594v2.

Christianson, Gale E. *Edwin Hubble: Mariner of the Nebulae* (Chicago: Univ. of Chicago Press, 1996).

———. "Edwin Hubble: Reluctant Cosmologist," *Historical Development of Modern Cosmology, ASP Conf. Proc.* 252 (2001).

Coles, Peter. "Einstein, Eddington and the 1919 Eclipse." arXiv:astro-ph/0102462v1.

Cowan, C. L. Jr., et al. "Detection of the Free Neutrino: A Confirmation," *Science* 124:103 (1956).

Cox, Brian. "A Journey to the Edge of Understanding," *Guardian*, June 30, 2008.

Dierickx, P., et al. "The VLT Primary Mirrors: Mirror Production and Measured Performance," *Proc. SPIE* 2871:385–92 (1997).

Dirac, Paul A. M. "Theory of Electrons and Positrons," Nobel Lecture. http://nobelprize.org/nobel_prizes/physics/laureates/1933/dirac-lecture.html.

Egret, Daniel, and André Heck, eds. *Harmonizing Cosmic Distance Scales in a Post-HIPPARCOS Era, ASP Conf. Series* 167 (1999).

Elst, Eric W. "Discovery and Rediscovery of Trojan Asteroids," *Earth, Moon, and Planets* 71:3, 275–77 (1995).

Esler, Karen J., Sue J. Milton, and W. R. J. Dean, eds. *Karoo Veld: Ecology and Management* (Pretoria: Briza Press, 2006).

Everett, Marshall. *Exciting Experiences in the Japanese Russian War* (Whitefish, MT: Kessinger Publishing, 2005).

Frenk, Carlos S., et al. "Cold Dark Matter, the Structure of Galactic Haloes and the Origin of the Hubble Sequence," *Nature* 317:595–97 (1985).

Gannon, Martin J. *Understanding Global Cultures: Metaphorical Journeys Through 28 Nations, Clusters of Nations, and Continents*, 3d ed. (Thousand Oaks, CA: Sage Publications, 2003).

Gibbons, G. W., S. W. Hawking, and S.T.C. Siklos, eds. *The Very Early Universe* (Cambridge, U.K.: Cambridge Univ. Press, 1983).

Glanz, James. "Listen Closely: From Tiny Hum Came Big Bang," *New York Times*, Apr. 30, 2001.

Glines, C. V. , ed. *Polar Aviation* (New York: Franklin Watts, 1964).

Gonzalez-Garcia, M. C., and F. Halzen. "Gamma Ray Burst Neutrinos Probing Quantum Gravity." arXiv:hep-ph/0611359v1.

Greene, Brian. *The Elegant Universe: Superstrings, Hidden Dimensions, and the Quest for the Ultimate Theory* (New York: W. W. Norton, 2003).

Gribbin, John. *In Search of Superstrings: Symmetry, Membranes, and the Theory of Everything* (Cambridge, U.K.: Icon Books, 2007).

Hale, Robert. *Christ and the Universe: Teilhard de Chardin and the Cosmos* (Chicago: Franciscan Herald Press, 1973).

Harper, Eamon. "Getting a Bang Out of Gamow," *GW*, Spring 2000.

Harper, Eamon, W. C. Parke, and David Anderson, eds. "Reminiscences of George Gamow," *ASP Conf. Series* 129 (1997).

Hawking, Stephen. *A Brief History of Time* (New York: Bantam Dell: 1988).

Henson, Paul, and Donald J. Usner. *The Natural History of Big Sur* (Berkeley: Univ. of California Press, 1993).

Hooper, Dan, Dean Morgan, and Elizabeth Winstanley. "Probing Quantum Decoherence with High-Energy Neutrinos," *Phys. Lett. B* 609:3–4, 206–11 (2005).

Hubble, Edwin. "A Relation Between Distance and Radial Velocity Among Extra-Galactic Nebulae," *Proc. Nat. Acad. Sci.* 15:3, 168–73 (1929).

Hudgins, Sharon. *The Other Side of Russia: A Slice of Life in Siberia and the Russian Far East* (College Station: Texas A&M Univ. Press, 2003).

Isaacson, Walter. *Einstein: His Life and Universe* (New York: Simon & Schuster, 2007).

Johnson, George. *Miss Leavitt's Stars: The Untold Story of the Woman Who Discovered How to Measure the Universe* (New York: W. W. Norton, 2005).

Kachru, Shamit, et al. "De Sitter Vacua in String Theory." arXiv:hep-th/030124v2.

Kamat, Sharmila. "Extending the Sensitivity to the Detection of WIMP Dark Matter with an Improved Understanding of the Limiting Neutron Backgrounds," Ph.D. thesis, Department of Physics, Case Western Reserve University, 2004.

Kevles, Daniel J. *The Physicists: The History of a Scientific Community in Modern America* (Cambridge MA: Harvard Univ. Press, 1995).

Khyentse, Dilgo. *The Heart Treasure of the Enlightened Ones* (Boston: Shambhala, 1992).

Kiryluk, J., et al. "IceCube Performance with Artificial Light Sources: The Road to Cascade Analyses," *Proc. 30th Int. Cosmic Ray Conf.* 3:1233–36 (2008)

Komatsu, E., et al. "Non-Gaussianity as a Probe of the Physics of the Primordial Universe and the Astrophysics of the Low Redshift Universe." arXiv:astro-ph.CO/0902.4759v1.

Lidman, Chris. "Observing Distant Type IA Supernovae with the ESO VLT," *Messenger* 118, Dec. 2004.

Linde, Andrei. "Chaotic Inflation," *Phys. Lett. B* 129:3–4, 177–81 (1983).

——. "Eternally Existing Self-Reproducing Chaotic Inflationary Universe," *Phys. Lett. B* 175:4, 395–400 (1986).

——. "A New Inflationary Universe Scenario: A Possible Solution of the Horizon, Flatness, Homogeneity, Isotropy and Primordial Monopole Problems," *Phys. Lett. B* 108:6, 389–93 (1982).

Martin, Geoffrey Lee. *Hellbent for the Pole: An Insider's Account of the "Race to the South Pole," 1957–58* (Auckland: Random House, 2007).

Mather, John C. "From the Big Bang to the Nobel Prize and Beyond," Nobel Lecture. http://nobelprize.org/nobel_prizes/physics/laureates/2006/mather_lecture.pdf.

McCarthy, Terence, and Bruce Rubidge. *The Story of Earth & Life: A Southern African Perspective on a 4.6-Billion-Year Journey* (Cape Town: Struik, 2005).

Middleton, Nick. *Going to Extremes: Mud, Sweat and Frozen Tears* (London: Macmillan, 2003).

Morgan, Dean, et al. "Probing Lorentz Invariance Violation in Atmospheric Neutrino Oscillations with a Neutrino Telescope." arXiv.astro-ph/0705.1897v1.

Morris, R. J. "Absence of Wax Esters in Pelagic Lake Baikal Fauna," *Lipids* 18:149–50 (1983).

Murayama, Hitoshi. "The Origin of Neutrino Mass," *Physics World*, 35–39, May 2002.

Norman, Nick, and Gavin Whitfield. *Geological Journeys: A Traveller's Guide to South Africa's Rocks and Landforms* (Cape Town: Struik, 2006).

Osterbrock, Donald E., Ronald S. Brashear, and Joel A. Gwinn. "Self-Made Cosmologist: The Education of Edwin Hubble." *Proc. Edwin Hubble Centennial Symposium, ASP Conf. Series* 10 (1990).

Ostriker, J. P., P.J.E. Peebles, and A. Yahil. "The Size and Mass of Galaxies, and the Mass of the Universe," *Astrophys. Jour.* 193:L1–4 (1974).

Perlmutter, S., et al. "Measurements of Ω and Λ from 42 High-Redshift Supernovae," *Astrophys. Jour.* 517:565–86 (1999).

Randerson, James. "Father of the God Particle," *Guardian*, June 30, 2008.

Reber, Grote. "A Play Entitled the Beginning of Radio Astronomy," *Jour. Roy. Astron. Soc. Can.* 82:3, 93–106 (1988).

Reines, Frederick. "The Neutrino: From Poltergeist to Particle," Nobel Lecture. http://nobelprize.org/nobel_prizes/physics/laureates/1995/reines-lecture.pdf.

Ribordy, M. "The IceCube Cosmological Connection: Status and Prospects of the Polar Neutrino Observatory." arXiv:astro-ph/0805.3546v1.

Riess, Adam G., et al. "Observational Evidence from Supernovae for an Accelerating Universe and a Cosmological Constant," *Astron. Jour.* 116:1009–38 (1998).

Rubin, Vera C. "Galaxy Dynamics and the Mass Density of the Universe," *Proc. Nat. Acad. Sci.* 90:4814–21 (1993).

Saha, P., et al. "COSMOGRAIL: The COSmological MOnitoring of GRAvItational Lenses IV. Models of Prospective Time-Delay Lenses." arXiv:astro-ph/0601370v1.

Sandage, Allan. *The Mount Wilson Observatory: Breaking the Code of Cosmic Evolution. Centennial History of the Carnegie Institution of Washington*, vol. I (Cambridge, U.K.: Cambridge Univ. Press, 2006).

Sarazin, Marc, et al. "Seeing Is Believing: New Facts About the Evolution of Seeing on Paranal," *Messenger* 132, 11–17, June 2008.

Schindler, Kevin. "An Amazing Tale: Slipher and the Spectrograph," *Lowell Observer,* Online Newsletter, Spring 2003.

Schneider, P., J. Ehlers, and E. E. Falco. *Gravitational Lenses* (Heidelberg: Springer-Verlag, 1999).

Scott, Robert Falcon. *Scott's Last Expedition,* vol. I. gutenberg.org/etext/11579.

Shackleton, Ernest. *South: The* Endurance *Expedition* (New York: Penguin Classics, 2004).

Sime, Ruth Lewin. *Lise Meitner: A Life in Physics* (Berkeley: Univ. of California Press, 1997).

Singh, Simon. *Big Bang: The Origin of the Universe* (New York: HarperCollins/ Fourth Estate, 2005).

Sinsheimer, Robert L. *The Strands of a Life: The Science of DNA and the Art of Education* (Berkeley: Univ. of California Press, 1994).

Slipher, V. M. "Nebulae," *Proc. Amer. Phil. Soc.* 56:403–9 (1917).

———. "Spectrographic Observations of Nebulae," *Pop. Astron.* 23:21–24 (1915).

Smoot, George, and Keay Davidson. *Wrinkles in Time* (New York: William Morrow, 1994).

Smothers, Ronald. "Commemorating a Discovery in Radio Astronomy," *New York Times,* June 9, 1998.

Spufford, Francis, ed. *The Antarctic* (London: Granta Books, 2008).

Starr, Kevin. *The Dream Endures: California Enters the 1940s* (New York: Oxford Univ. Press, 1997).

Strassmeier, K. G., et al. "First Time-Series Optical Photometry from Antarctica." arXiv:astro-ph/0807.2970v1.

Turley, Charles. *The Voyages of Captain Scott: Retold from the Voyage of the* Discovery *and Scott's Last Expedition.* gutenberg.org/ebooks/6721.

Weinberg, Steven. "Anthropic Bound on the Cosmological Constant," *Phys. Rev. Lett.* 59:22, 2607–10 (1987).

———. "Living in the Multiverse." arXiv:hep-th/0511037v1.

Wheeler, John Archibald, and Kenneth Ford. *Geons, Black Holes & Quantum Foam: A Life in Physics* (New York: W. W. Norton, 1998).

Wilkes, R. Jeffrey, J. G. Learned, and P. W. Gorham. "Deep Ocean Neutrino Detector Development: Contributions by the DUMAND Project." http://www.phys.hawaii.edu/dmnd/dumacomp.html.

Wischnewski, R. "The BAIKAL Neutrino Experiment: From NT200 to NT200+." arXiv:astro-ph/0609743v1.

Wouk, Herman. *A Hole in Texas* (Boston: Little, Brown, 2004).

Yoon, Carol Kaesuk. "Clues to Redwoods' Mighty Growth Emerge in Fog," *New York Times,* Nov. 24, 1998.

ACKNOWLEDGMENTS

The idea for this book appeared amid the stalled pages of an unfinished novel. Ravi Singh, publisher of Penguin (India) and a friend, patiently listened to why I couldn't deliver the novel, while helping me sort out my thoughts on a nonfiction book. The idea gathered force when Peter Tallack, my agent, expressed unreserved enthusiasm for it. Tentative thoughts turned into a book proposal, thanks to his acumen and cheerleading.

Most important, he found me Amanda Cook, my editor at Houghton Mifflin Harcourt — a writer's dream. She took the raw chapters — which at times seemed like disparate stories — and helped shape them to read like a book, while simultaneously focusing on the nitty-gritty: words, sentences, paragraphs. She's thoughtful, astute, and encouraging. I can't thank her enough. Sara Lippincott cast a copyeditor's eagle eye over the entire book. I remain amazed by her zeal and professionalism. A big thanks to her too.

This book would have been impossible to research and write without the enormous cooperation that was extended to me everywhere I went. My sincere thanks to everyone who made time to talk to me; explained the science; shared their knowledge of the astonishing telescopes, instruments, and sites; and extended their warmth and friendship. They are too numerous to be named individually, but that doesn't lessen my appreciation. Still, a few special thanks are in order.

Paranal, Chile: ESO's Claus Madsen helped set up the trip and Valentina Rodriguez made my stay there an absolute joy.

Lake Baikal, Siberia: Ralf Wischnewski, Kolja Budnev, and Alexey

Kochanov helped with all the paperwork to get me to Lake Baikal and took care of me once I got there. Tamara Nikolayevna, the head cook in that Siberian outpost, conjured up vegetarian meals in a land of meat and potatoes.

Toyama, Japan: The Super-Kamiokande detector in the Japanese Alps, unfortunately, did not get its own chapter in this book. Nonetheless, my sincere thanks to Yoichiro Suzuki, Yasuo Takeuchi, Shigetaka Moriyama, Yoshinari Hayato, and Masayuki Nakahata for their hospitality and for a chance to see what must be the most exquisitely engineered underground laboratory in the world.

Geneva, Switzerland: My thanks to Peter Jenni and Fabiola Gianotti for everything to do with my visits to CERN. And to Francois Butin for the tour of ATLAS.

Soudan, Minnesota: Richard Gaitskell and Dan Bauer made the trip possible. Michael Dragowsky, Jim Beaty, and Bill Miller were generous with their time and company inside the mine.

Mauna Kea, Hawaii: Without Sandra Faber's enthusiastic support, this journey would have fallen way short. And to Laura Kinoshita for driving me up to the summit of Mauna Kea.

The Karoo, South Africa: Adrian Tiplady took care of every detail of this trip, calmly and sympathetically dealing with a last-minute cancellation from my end (thanks to the airline clerk at London's Heathrow Airport who pointed out, just before check-in, that I had run out of pages in my passport). When I eventually did make it to South Africa, Adrian drove me to the Karoo and back (just in time for him to catch South Africa's victory over England in the Rugby World Cup finals). My thanks also to Bernie Fanaroff for his good humor, generosity, and time.

Antarctica: No praise is enough for the U.S. National Science Foundation's Antarctic Artists and Writers Program. They make you work hard to get a grant, but once you do, you are in supremely capable hands. I'm grateful to Francis Halzen and Steve Barwick for recommending me for the grant. My thanks to NSF's Kim Silverman for hand-holding during the application process, and Patricia Jackson of the U.S. Antarctic Program for everything else once the grant came through. The staff at Raytheon Polar Services Company in Centennial, Colorado, was superb, as were the folks at the U.S. McMurdo Station and the Amundsen-Scott South Pole Station. And my apologies to the doctor at the South Pole: Panicked that I had a piece of glass stuck in my throat (don't ask), I woke her up just after she'd gone to bed exhausted from a sixteen-hour shift.

Mount Wilson, California: Thanks to Don Nicholson, a sprightly ninety-year-old, for his enthusiasm and for being a link to Edwin Hubble. To Father Robert Hale and Father Bruno Barnhart at the New Camaldoli Hermitage in Big Sur, California, for sharing their thoughts on the contemplative life.

Hanle Valley, Ladakh: I'm grateful to Tushar Prabhu for single-handedly arranging a visit to this hauntingly beautiful land.

I'm grateful to Phil Austin, Stuart Clark, Alka Hingorani, Jenny Hogan, Bob Holmes, Rob Irion, Ben Longstaff, Valentina Rodriguez, Bruce Rubidge, and Richard Webb for reading parts or all of the book and/or for their advice.

I'm especially obliged to the physicists who read the parts pertaining to their work and suggested tweaks and pointed out errors: Michael Dragowsky, Dan Bauer, Ralf Wischnewski, Chris Lidman, Jim Peebles, Brian Gerke, Sandra Faber, Jocelyn Bell Burnell, Joseph Taylor, Bernie Fanaroff, Adrian Tiplady, John Mitchell, Gary Hill, Dan Hooper, John Kelley, Francis Halzen, Peter Jenni, Fabiola Gianotti, Andrei Linde, and Tushar Prabhu. The book is better for their efforts. Any errors that remain are, of course, solely my responsibility.

I could not have written this book without the support of my family. A special thanks to my parents for always standing behind me no matter what I choose to do, and also to my sisters, brothers-in-law, niece, and nephews (including one who, when he was six, offered to read a chapter, in all seriousness).

And finally, to the security guard in Vladivostok, Russia, who saw me sitting forlorn in an empty airport—having arrived, mistakenly, a full four hours before my 11:00 A.M. flight to Japan—and bought me coffee from the vending machine with his own change and would not accept my ruble notes in return: Thank you.

INDEX